THE AURIFEROUS GRAVELS OF CALIFORNIA.

GEOLOGY OF THEIR OCCURRENCE AND METHODS OF THEIR EXPLOITATION.

By JOHN HAYS HAMMOND, E.M.

Prior to the gold excitement in California and Australia in the years 1848 and 1851, respectively, the estimated quantity of gold in circulation in the world was from $2,000,000,000 to $2,500,000,000. Since the commencement of this golden era, it has been computed that the world's gold circulating medium has been increased by the introduction of about $4,500,000,000.

The great part California has played in the addition to the world's wealth will be recognized when it is said that fully one quarter of this increase has been derived from the gold mines of that State. Of the entire gold production of California, not less than nine tenths has been yielded by the "auriferous gravels."

The total yield of the auriferous gravels of California would be represented by the value of a cube of pure gold having an edge of fourteen feet. Over $100,000,000 have been invested in the gravel mines of California.

The auriferous gravels may be divided into two classes:

First—The shallow or modern placers.

Second—Deep or ancient placers.

These terms are indicative of the very characteristic difference that exists between the two classes of placers.

THE SHALLOW OR MODERN PLACERS.

The signification of these terms will be apparent from an explanation of the origin and mode of occurrence of the placers. These placers are superficial accumulations of auriferous alluvions along the gulches, bars, flats, etc., and are designated gulch, bar, or flat diggings, according to their topographical position. The deposits along the modern rivers belong to the shallow placers.

The gold-bearing detritus of which they consist has been derived from the ancient placers or the quartz veins; in some instances, from both sources. Through the disintegrating and transporting power of the meteoric agencies, especially of running water, the material has been brought from the above sources, and redeposited in places of lower elevation, where it was discovered by the pioneer miners. Thus the assorting process has been carried, by natural agencies, one step farther than in the case of the deep or ancient placers, and two steps farther than in the formation of the gold-

---

The relative costs of working the various classes of gold deposits by methods applicable to the respective classes are:

1. Auriferous vein _____$3 to $10 per ton material treated.
2. Drift mining_____75 cents to $4 per ton material treated.
3. Miner's pan_____$5 to $8 per ton material treated.
4. Rocker_____$2 to $3 per ton material treated.
5. Sluices _____75 cents to $1 per ton material treated.
6. Hydraulic method_____1½ cents to 8 cents per ton material treated.

bearing quartz veins.   To this process of the higher concentration of the gold is due the extraordinary richness of the shallow placers.

From the shallow placers, and from the beds of the modern rivers, came nearly all the gold produced by California up to the year 1854, and indeed the larger percentage up to the year 1860.   The exhaustion of the gold gravels, capable of being worked with profit by such primitive means as were available at that time, led to the abandonment of the shallow diggings.

The unworked deposits, also, often occur below the drainage level.

The pan and cradle, especially the former, are still used for the purpose of "prospecting" or testing the gravel, and for cleaning up the sluices, batteries, etc., at all mines.   The accompanying illustrations of these articles suffice to describe them.

The batea is better than the pan for purposes of prospecting gravel or quartz.   It is used extensively in South America, Central America, and Southern Mexico.   It consists of a shallow wooden bowl, from twelve to twenty-four inches in diameter and of two to four inches in depth from the center depression to the level of the rim.   It should be colored black (by paraffine paint) to show the collected gold more readily.   The pan is, however, preferable to the batea for cleaning up the sluices, etc.

The flat character of many of these deposits preventing the attainment of the grade requisite for "sluicing" and of the "dump" for the disposition of the tailings, enhances the difficulty of working alluvions of this class. To overcome these difficulties mechanical contrivances of various designs (hydraulic elevators, etc.) have been introduced in some places with considerable success.

River mining by sluices, rockers, etc., is now almost entirely limited to the operations of the Chinese, several thousand of whom, scattered over the State, follow the business in a desultory way.   A few of the bars of the present rivers are still being worked by "wing dams," by which at periods of low water the river is diverted to one side so as to leave dry the gravel deposit.   "Chinese" pumps are used in this class of mining to drain the deep holes from which the water has not receded after the deflection of the river.

In several localities, Butte and Placer Counties principally, plans are being carried out to work the beds of rivers reputed to be rich by constructing head dams, by which the water is diverted into large flumes, which carry it below the point to which it is designed to extract the gravel.   This system involves considerable expense, and is usually attended with great uncertainty as to the success of the enterprise.

Dredges are used in a few localities where gravel occurs below the drainage level of the locality to elevate the gravel to a horizon sufficiently high to obtain the requisite grade for sluicing and for the dump.

Under similar topographical conditions "hydraulic elevators" are used. The modus operandi is as follows: The material piped from the bank is carried by the water into an excavation into the bedrock, at which point is placed the lower end, made flaring, of the pipe through which the gravel is to be raised to the sluices, which are set on the level of the upper end of the pipe.   A stream of water, under high pressure, is played through a nozzle into the lower end of the pipe just described, and the gravel washed by the swiftly ascending current upwards through the pipe to the point of efflux into the sluices.

This method materially aids the disintegration of the gravel, wherefore a shorter string of sluices than ordinarily used will suffice.   Artotype No. 4 shows the elevator used at the North Bloomfield Mine.   The capacity of this elevator is, according to Superintendent Radford, two thousand four

MODE OF USING PAN AND CRADLE.

BRITTON & REY, S.F.

1. GULCH WASHING.—The coarse rocks are piled to one side, while the fine dirt is shoveled into the narrow sluices seen in the picture.
2. TRUSS BRIDGE, for carrying flume.

hundred cubic yards per twenty-four hours. The gravel is elevated eighty-seven feet vertically. One thousand three hundred miner's inches of water are required under a pressure of five hundred and thirty feet. In addition to elevating the gravel, the eight hundred inches of water used in piping the two thousand four hundred cubic yards of gravel is also raised with the gravel.*

The gold-saving apparatus of the shallow placer mine—the pan, cradle, tom, sluice, etc.—are now objects of hardly more than historical interest. From their nature, as compared with the illimitable resources of the deep placers, the shallow diggings, as a source of gold, will be but of a transient character. The virtual exhaustion of the superficial gravel deposits gave origin to that "prospecting" for the source of gold which resulted in the discovery of the deep placers and of the quartz veins.

From these discoveries dates the inauguration of mining as a scientific and permanent industry in this State.

EFFECTS OF EROSION.

The erosive power of flowing water is strikingly manifest in the stupendous denudation that has been effected through its agency on the western slope of the Sierra Nevada Mountains. In a very recent geologic time the entire relief of its western slope has been changed. Mountain ridges that were hundreds, and indeed thousands of feet in height, have been utterly obliterated by erosive agencies, while the present elevated position of the deep gravel channels shows how energetic has been the operation of these elements.

The prevalent topographical features of the gravel epoch have been in many instances inverted, and that which constituted the prominences of that age has been, by the resistless erosive power of water, converted into the depressions of to-day.

The detrital accumulations of the gravel epoch thus stand out in bold relief in the topographical features of the western slope of the Sierras, and are everywhere conspicuous objects in the panorama.

As early as 1851, Mr. T. P. Tyson, writing upon the geology and industrial resources of California, noted the existence and described the character of the gravel formation he had seen on his trip (in 1848) through the central mining counties.

THEORIES OF GENESIS OF "ANCIENT RIVERS."

Other travelers followed Mr. Tyson, each with his "theory" as to the genesis of the auriferous gravel formation, and it may be of interest to review a few of the more widely entertained of these theories, especially as some of them have not yet been wholly eradicated from the popular mind.

The "marine theory," which referred to the formation of the gravel deposits to the action of the sea, was, for many years, the most popular of the hypotheses evolved to explain the phenomena connected with the auriferous gravels.

The palæontological evidence on this point is most unequivocal. The absence of massive fossils on the one hand and the presence of remains of

---

* Since writing the above I learn from Mr. L. L. Robinson, the President of the North Bloomfield Company, that the hydraulic elevator has been abandoned because of the great expense attending its operation. He thinks, however, that for heights up to about forty feet the elevator would work to advantage under favorable conditions, as to cost of water, etc. The company is making other experiments in this direction at present.

terrestrial life on the other hand, are not only a most conclusive refutation of the "marine theory," but a positive proof that these deposits were of a subaerial origin. Other considerations (the distribution of the deposits, the character of the component parts of the formation, the remains of what were unmistakable portions of the banks of ancient rivers—the rim-rock, etc.) impel one to the above expressed conclusion.

The "glacial" theory was one which, for awhile, had many advocates. It finally lost its prestige when geologists conversant with glacial phenomena began to investigate this subject. The absence of the "bowlder clay" of the glacial striations, and of the rounded hummocks (called "roches mountinnées") is negative evidence regarding the glacial theory. Per contra, the character of the bedding, the sub-angular and rounded bowlders, show that water, not ice, has been the transporting agent in the accumulations of the auriferous gravels.

At the time Whitney began his examinations the "fluviatile, or deep river theory" had already many adherents. But, with this theory, was unfortunately connected the blue lead theory; or more accurately speaking, the "fluviatile" theory was embodied in the latter, which received the support of many geologists.

This (blue lead) theory supposed the existence of an ancient river flowing from the northeast to the southwest, and having its general course parallel to the crest of the Sierra Nevada Mountains.

The theory derived its name from the bluish color of the gravel filling the channel; and the adherents of this theory profess to be able to identify and to coördinate with the blue lead, by means of this characteristic color, isolated bodies of gravel in various portions of the State. The detailed examination of the geographical positions of the old river channels and their outliers has shown that no river or system of rivers followed this course in the gravel epoch, and the very foundation of the theory was thus destroyed. Furthermore, to this theory, as Mr. Goodyear says, exists the insuperable objection that its acceptance involves the assumption of an uplift or elevation of the Sierra Nevada Mountains, subsequent to the gravel epoch, to account for the change in the present system of drainage with reference to the system of drainage existing during the gravel period. That no such oscillation of level has occurred since the deposition of the deep gravel is, in the writer's judgment, susceptible of proof by the study of the structural geology of the Sierras.

Furthermore, that there is no necessary connection between the bluish color of the gravels and the source from which the material has been derived, will be demonstrated in connection with the discussion of the materials filling the channels.

The blue lead theory, like the glacial and marine theories before it, has now been generally discarded by the investigators of the auriferous gravels, and there has been generally accepted the "fluviatile" theory in its essence, ascribing the origin of the gravel accumulations to the depositions of ancient rivers with courses similar to the present rivers. The character of the deposits show that there has been an intermittent action. The torrential period and the quiescent period are both shown in alternating heavy and light material of the channels. But while there is this unanimity of opinion as to the origin of the gravel accumulations, there still exists a difference of opinion regarding the geological age of the deposits.

GEOLOGICAL AGE OF THE ALLUVIONS.

Authorities differ as to the age of these deposits. Whitney refers them to the tertiary (pliocene) period, while Le Conte assigns them to the qua-

HYDRAULIC ELEVATOR. BLOOMFIELD MINE.

ternary age. The palæontological evidence probably favors Whitney's classification. It is certain that the predominant flora and fauna are tertiary, but it is not impossible, Le Conte insists, that this pliocene life has lingered into quaternary time.

The mastodon is distinctly pleistocene, and remains of mastodons have been found in the deep river gravels.

MATERIAL FILLING THE CHANNELS.

Lava.—This may be a true lava, brought into its present position by a lava flow, or, as more generally is the case, the term refers to lava of a fragmental and tufaceous character. Lava bowlders are more or less rounded, water-worn bowlders of volcanic origin, brought into their present position by running water. Volcanic breccia and volcanic conglomerates are likewise designated by the term "lava." The most common form of lava is the material of the "light-colored, fine-grained, homogeneous beds, resulting from the consolidation of the ashes and volcanic mud."

The lava of the tufaceous variety is sometimes very compact, and has a metallic ring when struck, resembling "clink-stone" (phonolite) in this respect. In this lava crystals of glassy feldspar often occur.

The lava is nearly always found "capping" (overlying) the gravel, but in some localities it occurs interstratified with the gravel.

Beautiful dentritic markings (called "photograph rocks") are often found covering the lava. These markings are usually mistaken for impressions of plants, but they are of inorganic origin, having been formed by percolating waters, carrying oxide of manganese and oxide of iron, which substances have assumed the dendritic or tree-like form.

Cement is a term of variable signification. In some districts it designates volcanic material of brecciated or conglomerated character. In other localities it refers to a quartz (non-volcanic rock) conglomerate cemented by oxide of iron.

Pipe-clay is more or less indurated clay, of non-volcanic origin, forming finely laminated beds. It generally occurs interstratified with beds of sand, etc.

Plants, infusorial earth, etc., imbedded in the clay and lava deposits, are often found; fragmental specimens of leaves of various kinds, carbonized wood, and a sort of lignite is found abundantly in the channels. There is an abundance in many localities of petrified wood, formed by the silicifying actions of percolating siliceous waters. The fossil flora of the gravel beds is different from that of the present time, and is referred to the miocene or to the pliocene epoch of the tertiary age.

Often intercalated in the deposits of white lava, are found beds of infusorial earth. This substance is the remains of the silicious shells of microscopic plants (diatoms).

This earth has a magnesian or chalky aspect, is very light, and does not effervesce or dissolve with acids. Diatomaceous deposits are common in the European tertiary.*

Gravel.—The term refers to the water-worn pebbles or bowlders which occur generally as a more or less compact conglomerate, immediately overlying the bedrock. There is considerable heterogeneity in the lithological character of the conglomerate, though pebbles or bowlders of a quartzose character often predominate. Usually subordinate to the quartzose pebbles or bowlders are others, which are sometimes foreign to the lithology of the district, and have been transported long distances.

* The lava, volcanic sediment, and pipe-clay but very rarely carry any gold, and the presence of the gold in these formations is entirely fortuitous.

In some channels, the gravel consists entirely of quartz, while in others, quartz is altogether absent.

The cementing or binding material is sometimes calcareous, but generally ferruginous and siliceous. The infiltrating waters carrying the siliceous and other cementing material have most probably come from above, and not, as Whitney suggests, from below. In most instances lava has been the material from which the transmitting agencies have been derived.

Color of Gravel.—The term "red gravel" is given to the brownish or reddish colored conglomerate which forms the top and overlies the blue gravel, which is the bluish, grayish, or greenish colored gravel nearest to the bedrock.

The gravels owe their color to the difference in degree of oxidation and hydration of the iron constituents which form the cementing or binding matter of the conglomerate. The "blue gravel" has very often iron in the form of finely divided iron pyrites, and to the presence of this mineral, and of some of the ferrous salts, most probably is due the peculiar color. The oxidation of the ferrous salts and the pyrites converts the iron into anhydrous sesquioxide, which imparts the red color to the red gravel. The addition of water changes the anhydrous to the hydrous sesquioxide, and the gravel assumes a hue more or less brown, depending upon the degree of hydration, and is seen as a variety of red gravel.

As has been shown, the color of the gravel is entirely independent of the source from which it has been derived.

Color of Gravel as Indication of Value.—That the blue gravel is rich gravel cannot be true other than in a relative sense. It is true that the blue gravel is almost always richer than the red gravel, but, *per se*, the color of the gravel is no criterion of its absolute value. The gravel is richer nearer the bedrock than in the upper portions of the deposits, because of its great specific gravity, etc., and the color of the gravel in the lower portions is blue, because the lower horizons are less exposed to the oxidizing influence than the upper horizons of the deposit, where the atmospheric agencies have had access and have changed the original blue color of the deposit to the red color now characteristic of top gravel. The gold occurs disseminated throughout the greater portions of the red and blue gravel. It is not uniformly distributed, but is finer as regards its texture, and less abundant in the top gravel than below in the blue gravel. The pay gravel is generally confined to the blue gravel, and in some mines, especially where the more expensive process of drifting is adopted, the "pay gravel" rarely extends more than three to five feet above the bedrock. Where a schistose rock forms the bedrock, the gold is often found included between the laminæ, several feet below the bedrock.

The quartz bowlders found in gravel mines often carry considerable gold. At Polar Star Mine, Dutch Flat, a white quartz bowlder was found which contained $5,760 worth of gold. The gold has undoubtedly been derived from the gold-bearing quartz veins so numerous along the western slope of the Sierras. By water it has been transported and deposited along with the detritus in preëxisting valleys, where now found. In the quartz veins the gold rarely exists of such large size as often found in the gravel deposits, and this would seem to be incompatible with the theory of its source as above given.

Formation of Nuggets.—To explain this seeming incongruity several theories have been advanced. Professor Whitney is of the opinion that the gold veins were richer nearer the surface in pliocene times than now, and maintains that gold veins get poorer as depth is attained. In the judgment of the writer, who has examined nearly all the mines of this State, there is no connection between the richness of the ore and the depth at which found. A more plausible explanation of the formation of the nuggets is, as has been suggested by Le Conte and other geologists, that their large size is often due to a chemical as well as a mechanical phenomenon. It seems probable that the gold has been re-dissolved and re-precipitated. By oxidation, the iron pyrites has been changed into the sulphate. Percolating solutions of the sulphate of iron dissolve the gold with which it comes in contact, and coming in contact later with organic or other reducing agencies, the sulphate of iron is changed by them to the sulphuret, depositing the gold as this change takes place. By the deposition of gold in some such manner, from solutions constantly in the same place, nuggets are formed.

WIDTH OF CHANNELS.

The width of channel worked in hydraulicking varies from one hundred and fifty to one thousand feet. The width of the deep rivers on top varies from a few hundred feet to several thousand. At Columbia Hill, Iowa Hill, and other places, the rivers are from one mile to a mile and a half wide.

In some places, where the gravel deposits have a great extent, laterally the width may be due to the confluence of two or more streams or tributaries.

Broad expanses of alluvions may also indicate the embouchures of the ancient rivers at these points.

The configuration of the western slope of the Sierra Nevada Mountains, of tertiary times, differs from that of to-day in one important respect: in tertiary times the slope was more uniform and more gently undulating at the valleys, through which flowed the pliocene rivers, which were broader and shallower than those of to-day.

As a consequence of this, we find broad pliocene rivers as contrasted with the more narrow and torrential streams of our times. As Whitney says: "The gravel rivers, when choked with debris, had room to make for themselves new channels to one side or the other. They refused to be confined within fixed limits, and thus wear down one narrow channel, because the volume of water was too great."

This shifting of courses has given the ancient rivers their present labyrinthine aspect, and has complicated the study of their systems. The rivers are generally broadest at their bends, owing to the shifting of their courses, as the banks on the concave side wear away.

The uniformity that is generally considered as a characteristic feature of the distribution of the gold in gravel channels rarely exists. On the other hand there is almost always a great fluctuation in the value of gravel taken across the stream and likewise along the course of the stream. The larger channels are usually less "spotted" as regards the value of the gravel than the smaller channels.

In drift mining this tendency is well marked in the occurrence of the "pay leads," to which the extraction of gravel is restricted. These leads are usually near the lowest depression of the channel, but often make to one or the other side and have most sinuous courses and varying widths. The width of these pay leads is determined primarily by the cost of mining. Where low grade gravel can be profitably worked, the width of the pay lead obviously would be greater than where a more costly system prevails.

Pay leads one hundred to one hundred and fifty feet are regarded as of about the average width. Where rich gravel, $5 to $8, is the minimum grade worked, such leads are often only from fifty to seventy-five feet wide.

Where gravel of $2 to $4 is being mined, the width of the pay lead may often reach three hundred to four hundred feet.

GRADE OF THE RIVERS.

As with the rivers of the present day, the grade of the pliocene rivers is very irregular, varying from five feet to two hundred and fifty feet or more per mile. (In some of the smaller tributary channels, the grade is four hundred feet and upwards per mile.) In common with the rivers of to-day, they have falls, rapids, etc., and exhibit all other characteristic fluviatile phenomena. The direction of flow of the pliocene rivers is indicated by differences in level between points on the bedrock, and also, where settling has not occurred, by the bedding of the deposit. The smaller ends of rocks, it is generally believed, point down stream.

As compared with their modern representatives, the pliocene rivers had less grade in their upper portions and more in their lower courses. The gradients of the pliocene rivers were also more uniform, but even in some of the larger channels the grade increases or decreases rapidly within a short distance. In the Bloomfield Channel, within a thousand feet, the grade decreased from one hundred and fifty feet (its average grade) to fifty feet per mile.

DEPTH OF GRAVEL.

The depth of the gravel deposit (*i. e.*, the height of the bank as it now exists) is exceedingly variable, not only as between the different mines, but in the same mine, owing to the erosion of the surface, which has given rise to superficial inequalities.

In some places, as at Table Mountain, Tuolumne County (see section of this deposit), the gravel is not more than two feet in thickness. In other places the gravel deposit attains, it is stated, a depth of six hundred feet, as at Columbia Hill, Nevada County.

DISTRIBUTION OF GOLD.

The paragenesis of the gold in the auriferous gravels of California differs from that in similar formations elsewhere, in the comparative paucity of the associated minerals.

Zircon, magnetic pyrites (chiefly in the form of sand), and garnets are the most abundant minerals accompanying the gold; but platinum, iridosmine, rutile, epidote, diamonds, chronite, topaz, cassiterite, and other minerals, also occur. The gold appears in size from minute particles (finest flour) to large nuggets* weighing several pounds. Gold from size of flax seed to melon seed is, as a general thing, considered coarse gold. Gold from top gravel is textually fine, and also of high grade, the coarser gold being nearer the bedrock. The gold usually is of a flattened character, well rounded on its edges. Occasionally the gold is rough and but little water-worn, and frequently still in its matrix. Gold found in upper gravel often has preserved its crystalline form. Rusty gold is often found in the alluvions. The rusty character of the gold is probably due to a coating of silica and sesquioxide of iron. It is not readily, indeed at times not at all amalgamable. Such gold, when saved, is recovered entirely by rea-

---

* The largest mass of gold found in California, within the knowledge of the writer, was extracted from the vein in the Morgan Mine, Calaveras County. It weighed one hundred and ninety-five pounds, and was worth $43,534. One or two nuggets worth from $25,000 to $30,000 each are said to have been found in gravel deposits of this State. There are many well authenticated finds of nuggets ranging from $5,000 to $10,000 in value. Large nuggets rarely occur in the large hydraulic mines.

son of its specific gravity. The richest gravel, as a rule, occurs on and near the bedrock. Sometimes, however, equally rich gravel is found in the upper horizons of the channels, notably on benches occurring on the inner side of a bend in the course of a channel. At Cherokee, Butte County, immediately overlying the bedrock, is a coarse, strongly cemented conglomerate, of a bluish color, carrying but little quartz. The depth of this conglomerate will vary from ten to fifty feet. On top of this conglomerate is a layer of soft blue gravel, varying from one to ten feet in thickness. This layer carries gold quite freely. Overlying this soft blue gravel lies a stratum *sui generis*, called "rotten bowlders," which consist of decomposed quartz, carrying gold varying from $2 to $5 a cubic yard. Above this stratum lies a stratum of quicksand of unknown thickness. At the present time, on top of this stratum of quicksand lies a stratum of hard, tenaceous pipe-clay over three hundred feet in thickness. On top of this pipe-clay is a deposit of basalt one hundred to five hundred feet in thickness. Owing to the large quantity of barren superincumbent material, hydraulic mining has been abandoned. The company purpose to introduce drift mining. The old washings of surface ground near Malakoff, Nevada County, from 1870 to 1874, is estimated at about three million two hundred and fifty thousand cubic yards, the yield of which was about $2\frac{9}{10}$ cents per cubic yard. The Bloomfield Company, from November 29, 1876, to October 13, 1877, washed one million five hundred and ninety-one thousand seven hundred and thirty cubic yards of top gravel, which yielded $3\frac{8}{10}$ cents per cubic yard. During the same period that company washed seven hundred and two thousand two hundred cubic yards of bottom gravel, which yielded $32\frac{9}{10}$ cents per cubic yard. The bottom gravel extended from bedrock to a height of sixty-five feet. All the gravel above that horizon was considered as top gravel.

The depth of the top gravel varied from a few feet to over two hundred feet.

The steep bedrock is more favorable for rich gravel than where the bedrock is flat, but often the gravel is poor where the channel is narrow and steep.

Gravel that yields remunerative results by the drift process has been worked in some instances from sixty to one hundred and fifty feet above bedrock. The coarser gravel ("heaviest wash") is the richest. The light "wash" and material of a sandy character, is usually poor in gold. The top gravel of hydraulic mines rarely pays for piping, unless worked in conjunction with the underlying richer auriferous deposits. The gold is often found in the grass roots of the vegetation covering the gravel deposits, but not in considerable quantities. The top gravel of a deposit is usually richer than that lying a foot or two below the surface, owing to the concentration of the gold superficially. In schistose rocks, as well as in decomposed granitic rocks, the gold is often found in the bedrock at a depth of from one to five feet below the surface. In drift mining, the bedrock is stripped to recover this gold. In hydraulic mining, bedrock of this character, when not too hard, is piped. When necessary, bedrock is blasted before being piped.

BEDROCK.

Bedrock is the rock which formed the bed over which the ancient river flowed. Slate is the common bedrock, though granite forms the bedrock in some localities. In different parts of its course the bedrock of a channel may change in character. The slate varies greatly in its lithological character. Chloritic, talcose, serpentine, and siliceous slates predominate. Schistose rocks are regarded as more favorable bedrock than granitic.

Where the stream runs with the stratification of the bedrock, the gravel is usually richer than where it cuts across the stratification. In some deposits strata of indurated clay, or of compactly cemented gravel, form what is called a false bedrock. This bedrock often occupies an horizon considerably above the true bedrock. Sometimes the conditions of drainage necessitate the adoption of this stratum as the working bedrock of the mine, and operations are conducted with reference to this false bedrock, in lieu of the true bedrock.

YIELD OF CHANNELS.

The aggregate length of the ancient channels has been estimated at four hundred miles. This does not include the so called cement channels, which are but of subordinate economic importance. The yield, per mile, of channels of the average character is, at a low estimate, from $2,000,000 to $3,000,000. Good channels for drifting yield from $100 to $500 per lineal foot of the stratum extracted. From one fifth to one half of the gold in the channels usually may be obtained from drifting where the bedrock gravel is accessible; the lowest stratum of gravel, four to eight feet in depth, carries this portion of the aggregate gold contents in the channel.

SECTIONS OF GRAVEL DEPOSITS.

Section across Table Mountain (Tuolumne County), after Whitney.

1. Top. Basalt, overlying andesitic cement, sixty feet (forming a bluff sixty feet high).
2. Bluish soft sandstone, seventy feet.
3. Whitish clays, ten feet.
4. Auriferous gravels, two feet.
5. Bedrock slate.

Section through Shaft 1, Malakoff (North Bloomfield Gravel Mining Company).

Depth of eight to ten feet, soil.
To depth one hundred feet, gravel.
At one hundred feet, streaks of clay mixed with the gravel for a few feet.
At one hundred and twenty feet, a foot or two of sand.
At one hundred and thirty feet, much quartz; pipe-clay disappeared.
At one hundred and forty feet, gravel became more firmly cemented.
At one hundred and seventy-five feet, all the gravel well cemented.
At one hundred and ninety-five feet, thin strata of sand for a few feet.
Then to two hundred and five feet, more cemented conglomerate.

Section at La Grange Mine (Stanislaus County).

1. Top. Red soil colored by Fe_2O_3.
2. Coarse red gravel (containing granite pebbles, etc.).
3. Red cement (hard pan) gravel cemented by Fe_2O_3.
4. White siliceous clay.
5. Red cement (as No. 3).
6. Sand with pebbles.
7. Loose yellow sand.
8. Dark-colored gravel, containing debris of granite, slate, diorite, serpentine, etc., with some quartz.

It will be observed that quartz forms but a small part of the gravel at La Grange, which is indeed composed almost entirely of granite, diorite, etc. This is quite anomalous in the geology of the auriferous gravels.

The above sections will show how great a diversity exists in the relative depth or thickness of the material filling the channels. In some localities, the gravels are capped (overlaid) by two hundred to one thousand feet of "lava," "pipe-clay," sand, etc. In other places, as at Bloomfield, etc., where hydraulic mining was most successfully conducted, the volcanic

formation is entirely subordinate, and comparatively little of the non-paying superincumbent material has to be "piped off" before the pay gravel
is reached. The "lava" capping in Placer and Nevada County is not the
"lava" flow which caps the gravel formations at places in Tuolumne,
Butte, and other counties, but it is of a sedimentary origin.

In some channels, where the grade is very steep, there was no deposit of
gravel. A subsequent deposition of volcanic sediment covered the bedrock. This occurrence is to be seen in the Black Cañon Channel.

SKETCH 1.

CROSS SECTION OF SECRET HILL AND CANADA HILL CHANNELS.

This sketch shows the *elevated* position of this ancient river system, with
respect to the modern drainage system (North Fork of American River), and
the sea level. Some gravel deposits of this class occupy altitudes even
higher than the Canada Hill system. This deposit has been drifted at
several points, and also hydraulicked where the volcanic capping had been
denuded.

E. represents an isolated bench of gravel about one half mile long and
eight hundred feet wide. The grade of this channel is very steep, being
about six feet per one hundred feet. Channel is about three hundred feet
wide at gravel line.

SKETCH 2.

CROSS SECTION OF RIDGE BETWEEN NORTH AND MIDDLE FORKS OF THE AMERICAN RIVER.

This sketch shows where the deep river system is *below* the drainage
level of the district. The diversion of the ancient stream into other channels was caused by the volcanic flow. The submerged position of the
deep river channels with respect to the present river systems is not usual
in the upper portions of their courses. Near Oroville, the Feather River
passes over a submerged ancient channel.

SKETCH 3.

LONGITUDINAL SECTION OF DAMASCUS CHANNEL AND CROSS SECTION OF THE RED POINT CHANNEL.

This sketch shows the older (Damascus) channel, which has been cut
through by the younger (Red Point) channel. The gravels filling these
channels have been derived respectively from different sources. The gravel
of the Damascus Channel is white, and is almost entirely composed of
quartz pebbles and bowlders, while that of the Red Point Channel is black,
and consists of a slate, very siliceous and highly metamorphosed.

SKETCH 4.

CROSS SECTION OF DAMASCUS AND BOB LEWIS CHANNELS.

Dotted lines, M. N. and M'. N'. show the original channel. These rims
have been washed away since the "volcanic" deposit was found. F. F. F.
are benches of gravel left on the west rim of the ancient channel. There
were, probably, corresponding benches on the east rim, which, together

with the rim, have been washed off and obliterated. Subsequent to the deposition of the gravel in the Damascus Channel system (C. and F. F. F.) a new channel (E.) has crossed and scoured out the older channel (C.). The gravel of the Damascus Channel is white, while that of the Bob Lewis Channel is black. The dotted lines M. N. and M'. N'. represent the original Damascus Channel's rims. The east rim of the original channel has been entirely carried away by the younger (Bob Lewis) channel.

SKETCH 5.

CROSS SECTION OF RED POINT CHANNEL AT HOG'S BACK AND BLACK CAÑON.

The Black Cañon Channel has a grade of eight feet to the one hundred in places. The space included within the dotted lines N. M. Y. M'. N'. has been obviously denuded since the deposition of the volcanic cappings. M. represents the vertical depth of the erosion subsequent to the volcanic formation. The vertical depth at this point is from one thousand eight hundred to two thousand feet. The dotted lines N. M. R. and N'. M'. R'. are the rims of the ancient channel. These rims were eroded subsequent to the deposition of the volcanic cappings. The harder capping has resisted erosion, while the softer slate rims have been denuded. Dotted line M. M. represents level of channel at time of lava flow.

SKETCH 6.

CROSS SECTION OF RED POINT AND DAMM CHANNELS.

E. is mixed quartzose (white) gravel on rim. This was left by some ancient channel which was subsequently scoured out by Red Point Channel (black gravel channel). Dotted line N. M. denotes the probable original west rim of the ancient channel, and N'. M'. the probable east rim. There have been, probably, three channels of different ages in this locality. The quartz channel, of which E. is a remnant, has been scoured out and obliterated by the younger black slate channel (Red Point Channel). The east rim, N'. M'., of the valley, through which the ancient Red Point River flowed, has been partly eroded by the youngest river of the three systems; i. e., the Damm Channel.

SKETCHES 7 AND 8.

LONGITUDINAL AND CROSS SECTIONS OF IOWA HILL CHANNEL SYSTEM.

The gravel channel, as illustrated in this sketch, has been entirely carried off in places through erosion by the North Fork of the American River, and by the waters of the Indian Cañon. The volcanic capping (volcanic sediment consolidated), which originally covered the channel, has been denuded at Iowa Hill and at Wisconsin Hill. At those points the gravel has been profitably hydraulicked, while the gravel which underlies the Morning Star Hill is not susceptible of being hydraulicked on account of the superincumbent volcanic capping. Where this capping was not more than fifty feet deep hydraulicking paid. The lower stratum of blue cemented gravel has been extensively drifted. The width of this channel shown in plate 8 (cross section) is nearly two miles.

NOTE.—These sketches are not made to scale.

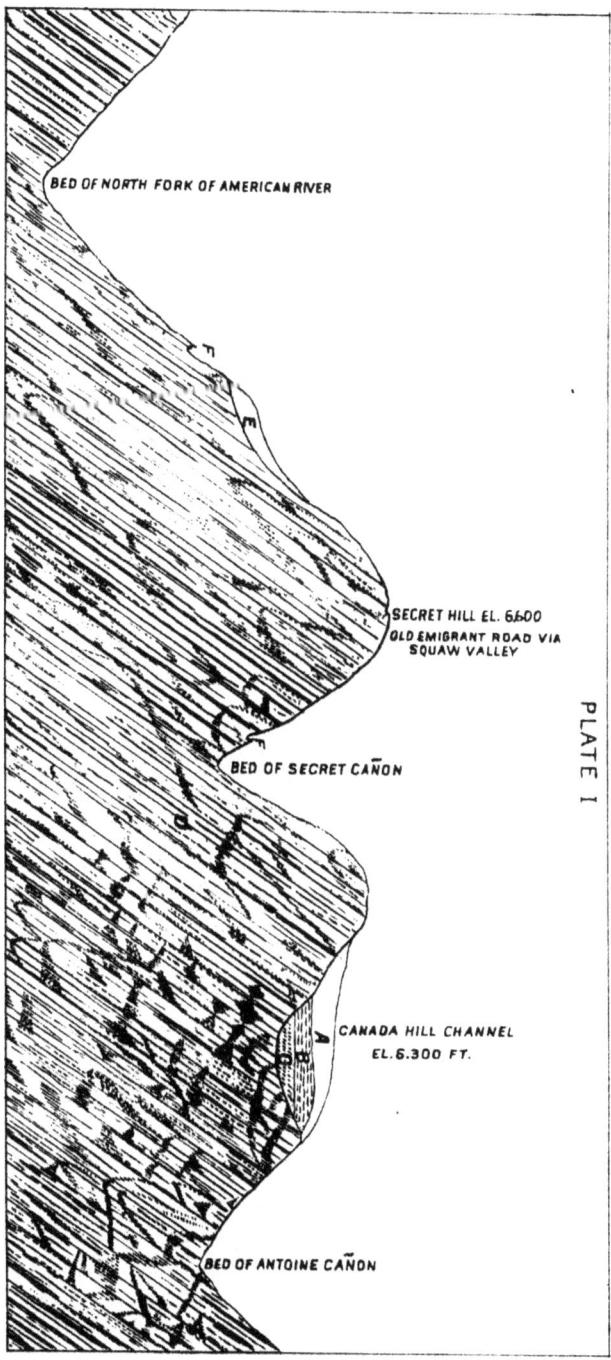

CROSS-SECTION OF SECRET HILL AND CANADA HILL CHANNEL. SKETCH (AFTER HOBSON).

PLATE I

BED OF NORTH FORK OF AMERICAN RIVER

F
E

SECRET HILL EL. 6,600
OLD EMIGRANT ROAD VIA SQUAW VALLEY

BED OF SECRET CAÑON

CANADA HILL CHANNEL
EL. 6,300 FT.

BED OF ANTOINE CAÑON

A. Volcanic Cement Capping, 40′ to 150′ deep.
B. Cemented White Siliceous Sediment 40′ deep.
C. Angular Gravel, quartzose, said to contain from $1 to $5 per ton.

D. Slate Bed-rock, gashed with numerous veins of quartz.
E. Bench of Gravel.
F. Iowa Hill Canal.

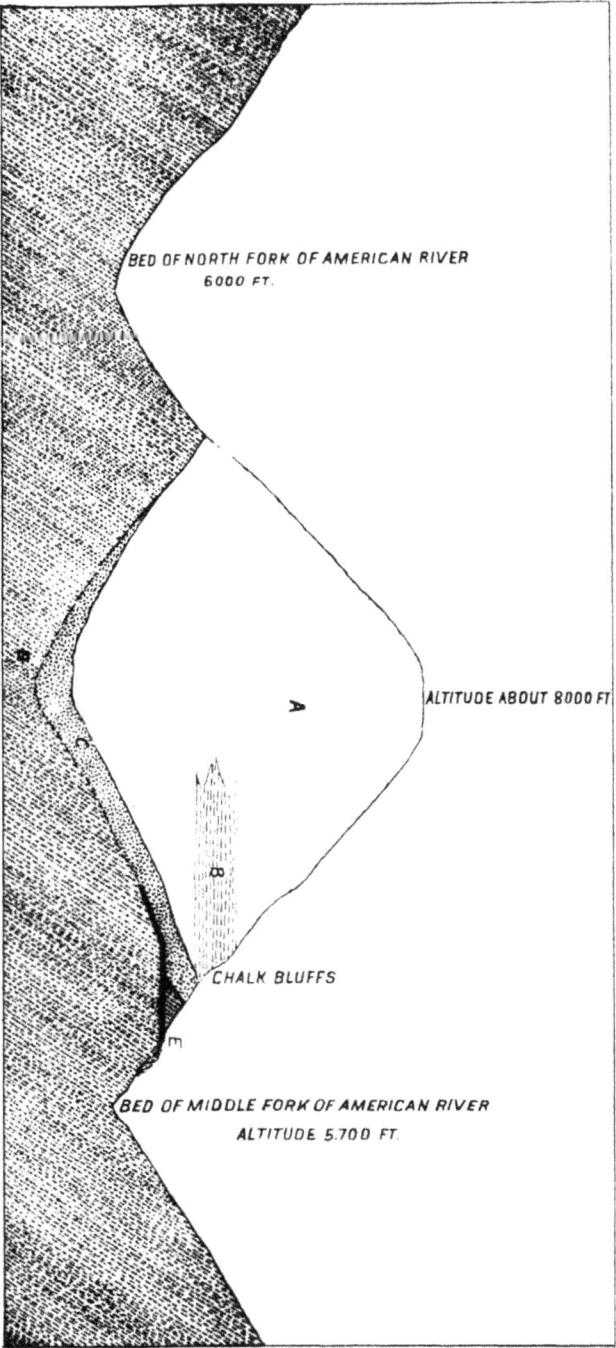

PLATE 2

CROSS-SECTION OF RIDGE BETWEEN NORTH AND MIDDLE FORKS OF AMERICAN RIVER.

BED OF NORTH FORK OF AMERICAN RIVER
6000 FT.

ALTITUDE ABOUT 8000 FT

CHALK BLUFFS

BED OF MIDDLE FORK OF AMERICAN RIVER
ALTITUDE 5.700 FT.

A. Volcanic Capping.
B. Cemented White Siliceous Sediment.
C. Gold-bearing Gravel.

D. Granite Bed-rock.
E. Prospect Tunnel and Slope (inclined shaft) sunk on the rim to work the gravel deposit.

PLATE 3

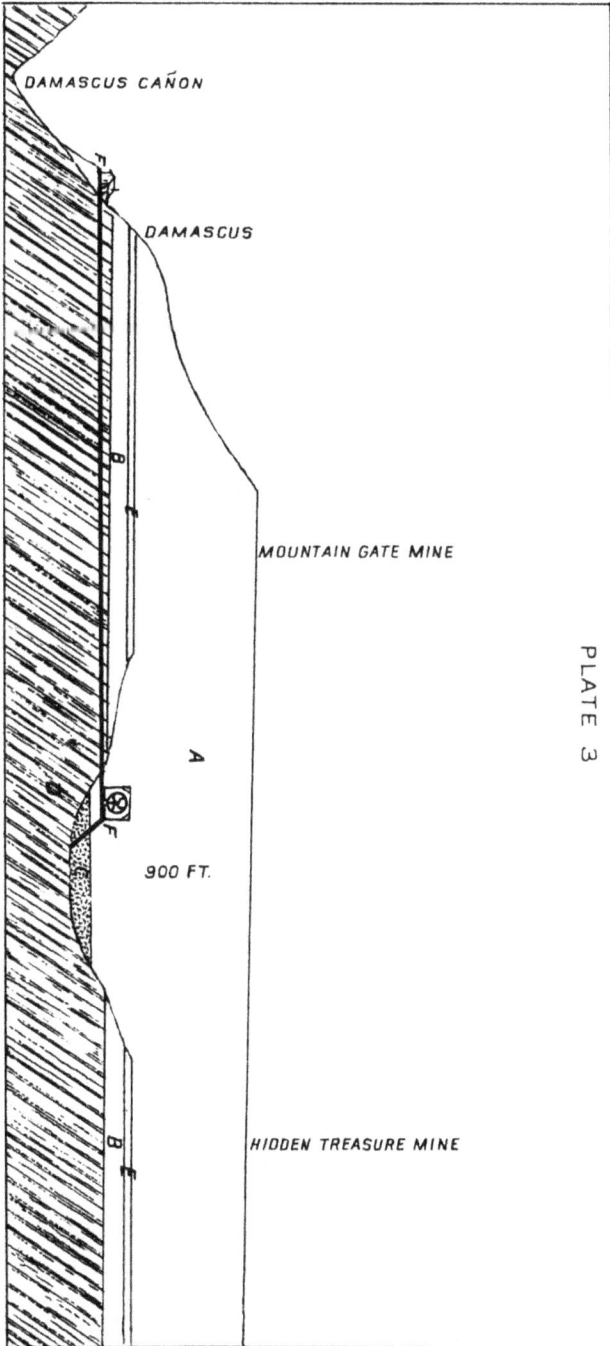

LONGITUDINAL SECTION OF DAMASCUS CHANNEL, AND CROSS-SECTION OF RED POINT CHANNEL.

DAMASCUS CAÑON

DAMASCUS

MOUNTAIN GATE MINE

900 FT.

HIDDEN TREASURE MINE

A. Volcanic Capping.
B. Damascus Channel (Grave composed almost entirely of white quartz.)
C. Red Point Channel (Gravel composed almost entirely of black, highly siliceous slate.)

D. Slate Bed-rock.
E. Pipe-Clay.
F. Working Tunnel and Incline.

PLATE 4

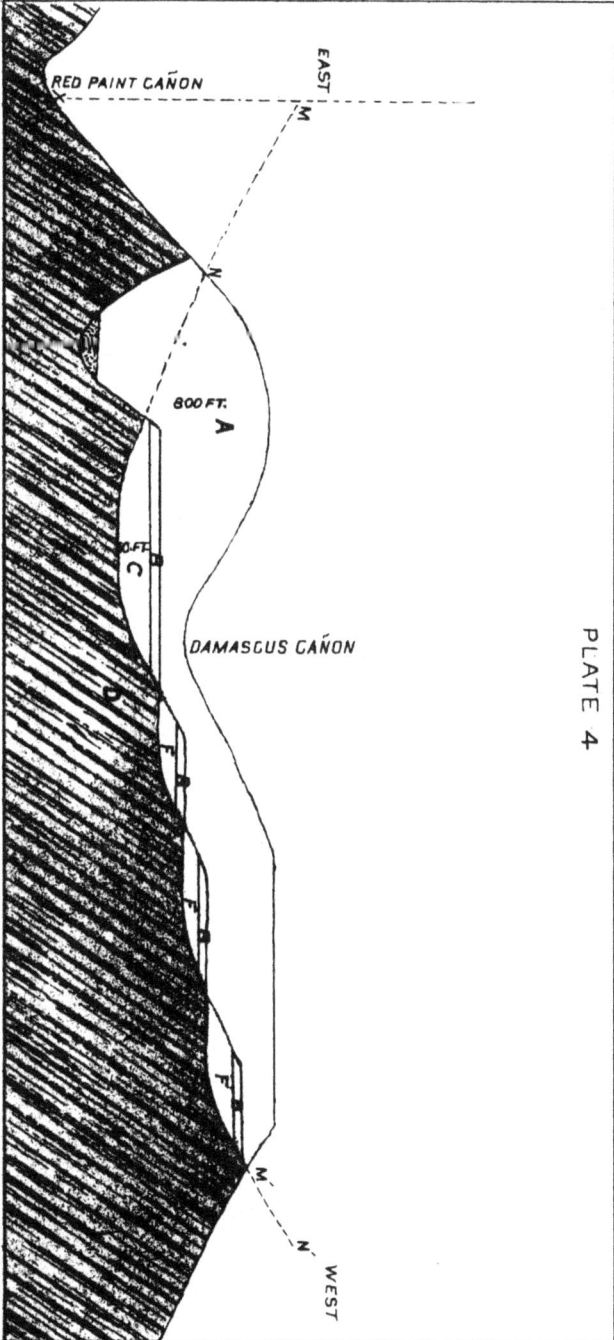

CROSS-SECTION OF DAMASCUS AND BOB LEWIS CHANNELS.

A. Volcanic Capping.
B. Pipe-Clay.
C. Damascus Channel, white quartz gravel.

D. Slate Bed-rock.
E. Bob Lewis Channel, dark gravel.
F. Rim benches of quartz gravel.

E is 150′ wide.

C is 1000′ wide, and has been worked to width of 300′.

RED PAINT CAÑON

EAST

800 FT.

DAMASCUS CAÑON

WEST

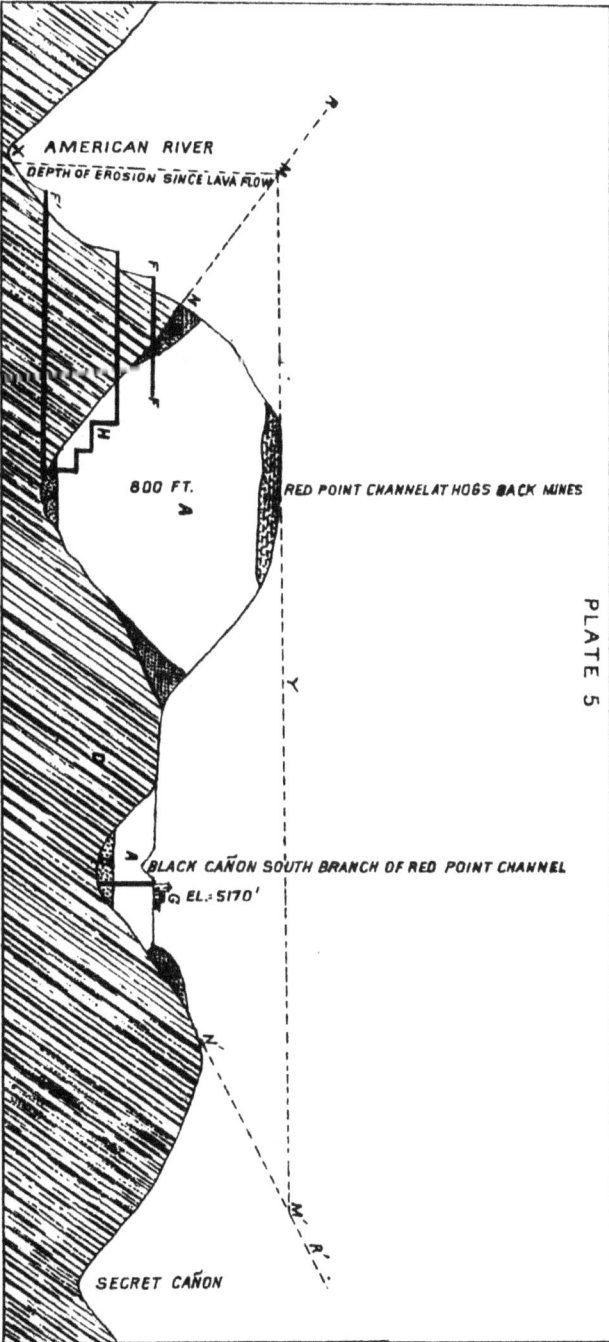

CROSS-SECTION OF RED POINT, HOGS BACK AND BLACK CANON CHANNELS.

A. Volcanic Capping.
B. Gold-bearing Gravel Channels.
C. Mixed Gravel on Rim.
D. Slate Bed-rock.

F. Prospect Tunnels.
F¹. Tunnel being run to bottom gravel deposit.
G. Black Canon Shaft and Hoisting Works.
H. Prospect Winzes to strike Gravel.

PLATE 5

AMERICAN RIVER

DEPTH OF EROSION SINCE LAVA FLOW

800 FT.

RED POINT CHANNEL AT HOGS BACK MINES

BLACK CAÑON SOUTH BRANCH OF RED POINT CHANNEL

EL. 5170'

SECRET CAÑON

PLATE 6

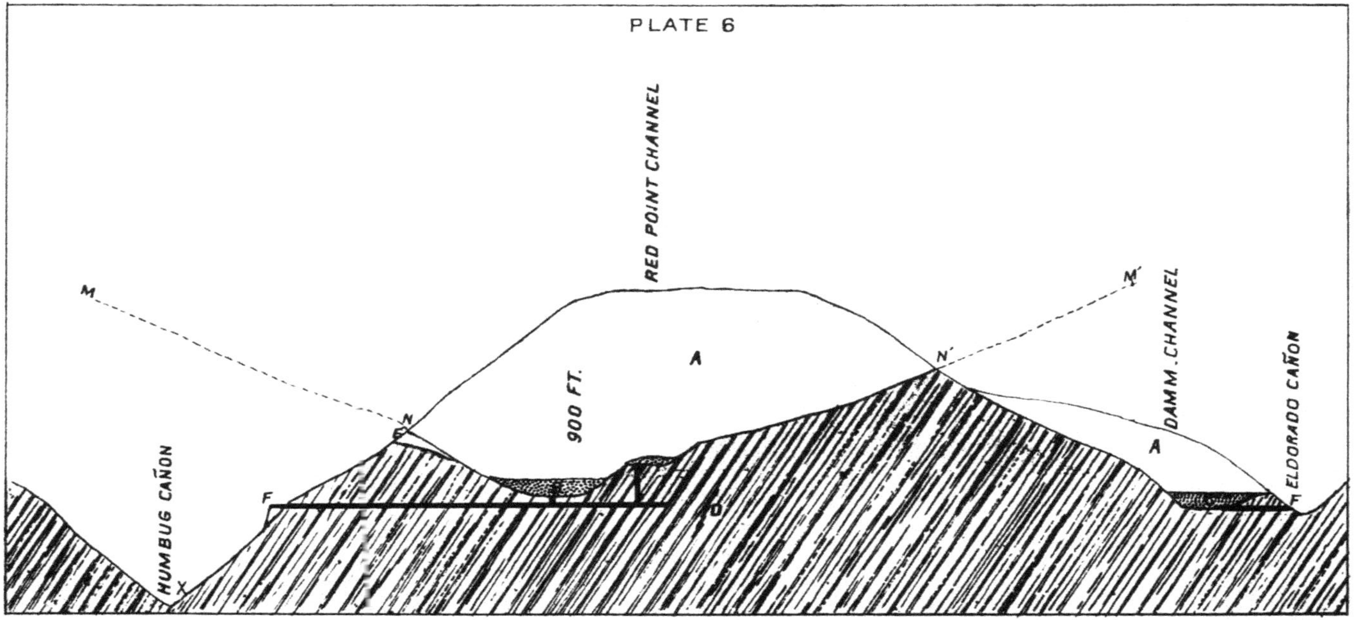
M
N
E
F
RED POINT CHANNEL
900 FT.
A
N'
M'
DAMM. CHANNEL
ELDORADO CAÑON
HUMBUG CAÑON
X
D

CROSS-SECTION OF RED POINT AND DAMM CHANNELS.

A. Volcanic Capping.
B. Red Point Channel.
C. Damm Channel.
D. Slate Bed-rock.
E. White Quartz Gravel on Rim.
F. Working Tunnels.

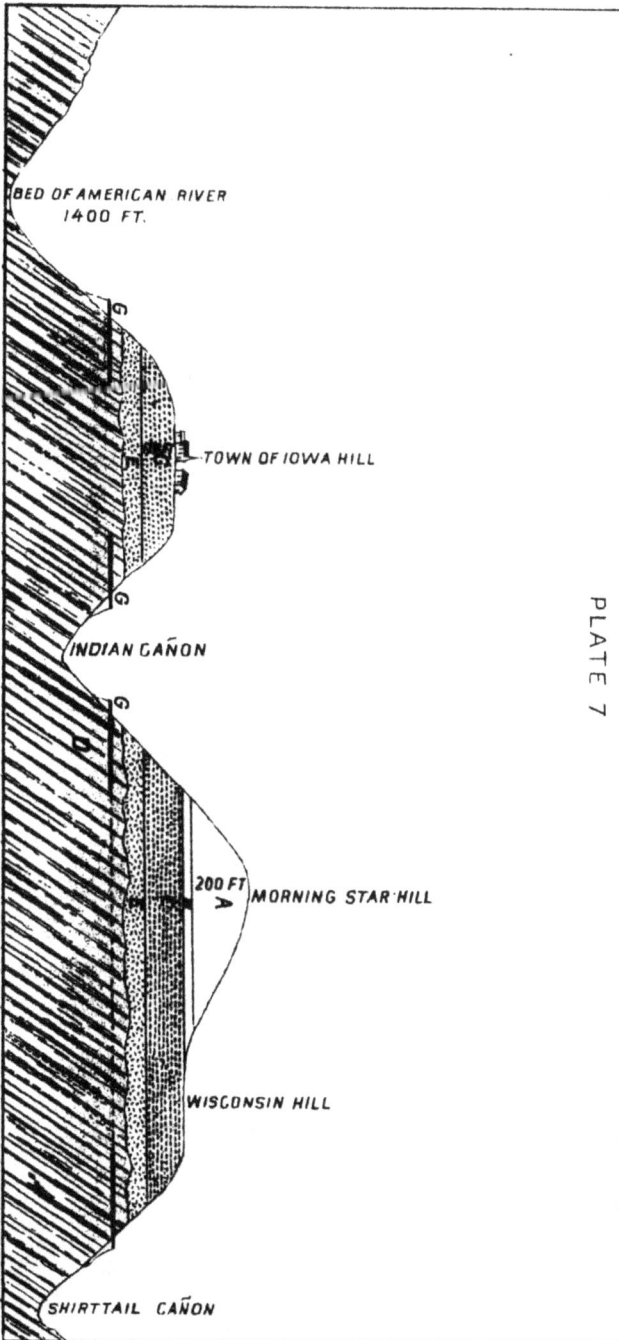

PLATE 7

LONGITUDINAL SECTION OF IOWA HILL CHANNEL SYSTEM

A. Volcanic Capping.
B. Pipe Clay.
C. Mixed Quartz Gravel, generally fine.

D. Slate Bed-rock.
E. Blue Gravel of Iowa Hill main Channel.
G. G. Tunnels

BED OF AMERICAN RIVER 1400 FT.

TOWN OF IOWA HILL

INDIAN CAÑON

200 FT MORNING STAR HILL

WISCONSIN HILL

SHIRTTAIL CAÑON

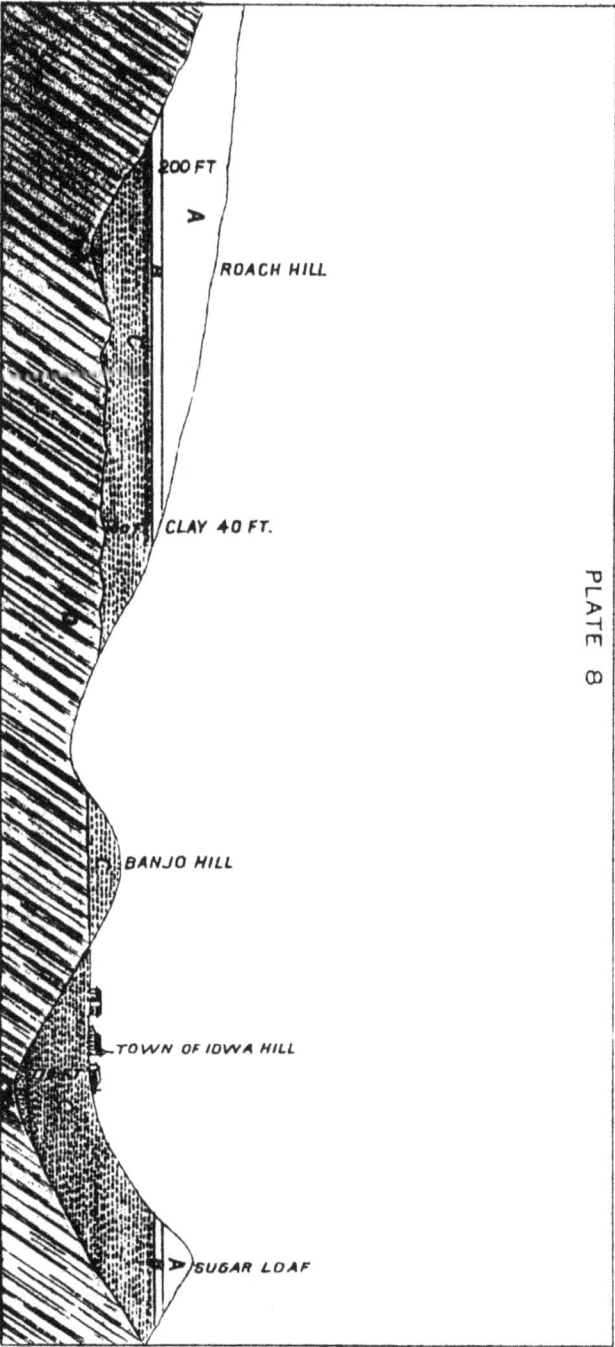

PLATE 8

CROSS-SECTION OF IOWA HILL CHANNEL SYSTEM.

200 FT

A

ROACH HILL

CLAY 40 FT.

BANJO HILL

TOWN OF IOWA HILL

A SUGAR LOAF

A. Volcanic Capping
B. Pipe Clay.
C. Mixed Quartz Gravel.
D. Underlying Slate Rock

E. Blue Channel, Wolverine Channel.
F. Main Blue Gravel, Iowa Hill Channel.
G. G. Tunnels.

METHODS OF MINING.

The mining of the deep placers may be divided into two classes, viz.: drift mining and hydraulic mining. As is well known, the hydraulic mining has been almost entirely restricted during the past six years by adverse legal decisions. This stoppage of hydraulic mining has given a great stimulus to drift mining, many of the enjoined hydraulic mines being now worked by drifting. But there are many gravel deposits that are peculiarly adapted to profitable hydraulic mining that are not susceptible of remunerative drift mining. In some mines the gold is more or less uniformly distributed throughout the gravel, and the lowest stratum is not rich enough to pay for the drifting, while hydraulicking can be profitably conducted.

The lithological character of the material filling the old river channels will determine which of the systems is to be pursued. Where a great depth of lava (as in the case of the Table Mountain formation, a section of which has been given) or other non-paying material is superimposed upon the auriferous pay gravel, drift mining must be resorted to. The exploitation of such large quantities of barren material would not be practicable, and by the system of drift mining this is obviated. In some places there is a concentration of the gold in a stratum immediately overlying the bed-rock, while the rest of the deposit carries an amount of gold too small to admit of profitably hydraulicking the entire deposit. When such is the case, recourse is had to drift mining. Again, lack of grade, lack of dump, or the absence of other conditions requisite for hydraulic mining, make it necessary to adopt drift mining instead of the hydraulic system.

In drift mining, the deposit is worked through a shaft or through a bed-rock tunnel. The latter system is preferable and is adopted where the topography of the district admits, inasmuch as the necessity of draining the mine by pumping and of elevating the gravel is obviated. Fortunately, the topographical features are often favorable to the working of the deposits by tunnels. To reach the gravel deposits, a bedrock tunnel is run from some neighboring cañon through the rim-rock to a point beneath the gravel at which exploitation is to begin. *In drift mining, where worked through tunnels, it is vitally important to determine the approximate elevation of the gravel deposits, so as to run the tunnels sufficiently below the gravel to admit of working the deposit by gravitation. When the tunnel is run above the bedrock (except in cases where a false bedrock is adopted as working level), the deposit cannot be worked without elevating the water and gravel.

This is a costly and rarely practicable system of mining. Many disastrous mining failures have attended the faulty location of a bedrock tunnel. The bedrock tunnels are sometimes more than a mile in length. These tunnels cost from $5 to $40 per running foot, $12 to $16 being about the average, depending upon their dimensions, and chiefly upon the hardness of the ground. Power drills are generally used in running long tunnels. While in rock of medium hardness there is no economy in the use of these drills; in effecting a saving of time they are valuable. On the other hand, where the rock is exceptionally hard, the use of air drills effects

---

* Within a few years, the projection of the ancient channel upon maps to distances of several miles beyond the nearest visible point has been correctly made by engineers, who have made a specialty of this class of engineering work. In this connection the delineation of the Red Point Channel (see sketches 3, 5, 6) and the determination of the best site for a tunnel to open up the gravel deposit, is deserving of note. This successful and very creditable work was done by Messrs. Ross E. Browne, John B. Hobson, and Charles Hoffman. (See article by Russell L. Dunn, in Report of State Mineralogist 1888.)

9 27

the saving both of money and time.   The dimensions of the tunnels, where run by hand, are generally about five feet wide and six and one half feet high.   Where run by air drills, they are often as wide as eight to ten feet, and from seven to eight feet in height.   Sometimes, to make speed, two drills are worked at the same face, in which case the width of the tunnel is from eight to ten feet.   From one hundred and fifty to four hundred feet of tunnel may be run monthly by using air drills.   The tunnel, where run in bedrock, but rarely requires much lagging and timbering.   When run in gravel, this is an expensive item of cost.   From the tunnel, upraises are made into the gravel deposit.   These upraises are one hundred and fifty to two hundred feet apart, and serve as chutes.   From the top of the upraises, gangways are run through the gravel, and the deposit is blocked out by cross drifts, and exploited in a manner similar to the working of horizontal seams of coal.   The gravel is usually worked to a height of five to eight feet.   Where the bedrock is soft, and of a laminated character, it is taken out to a depth of from one half foot to four feet, and washed with the gravel.   The large bowlders are picked out and stowed back in the excavated ground.   The gravel is wheeled, or brought by small cars, to the gravel chutes, leading to the bedrock tunnels.   Through these chutes it is dumped into the cars in the main tunnel.   These cars, where the tunnel is long, are drawn (six to ten constituting the train) by animals, or by locomotives,* and emptied automatically upon "dumping chairs" near the mouth of the tunnel.   The cars carry from one to two tons of gravel.   The best grade for the tunnel is three to four inches per hundred feet.

The gauge of the track is from sixteen to twenty-two inches.   T rails, weighing from twelve to thirty pounds, are generally used, but in some mines wooden rails, covered with strap iron one to two inches in width by one quarter to one half inch thick, are used.   T rails are unquestionably preferable.   Track of this character can be laid at from 50 to 60 cents per foot.   In the gangways through the gravel, the track is a temporary structure that can be conveniently moved from the abandoned gangways.

Outside of the tunnel the track divides into three branches.   One track leads to the waste dump, where the waste rock from the bedrock tunnel, etc., is dumped.   Another track leads to a small "prospect" bin, into which the gravel to be tested as to its richness is discharged.   The third (main) track leads to the main gravel dump.

EXTRACTION OF THE GOLD.

The gold is extracted either by:

First—Sluicing, when the gravel is " free," *i. e.*, not cemented.

Second—By the milling process, when the gravel is cemented.

The process of sluicing is essentially similar to the washing of the gravel in hydraulic mining.   The cars are run on to dumping chairs, which dump the gravel automatically.   The gravel falls into a wooden bin, heavily boarded, and sometimes covered with light sheet iron.   At the bottom of the bin is the head of the line of sluices, into which the gravel is directed by the slope given to the sides of the bin.   This structure is generally covered over to protect the men from the inclemency of the weather.   The water is introduced by a pipe connected with the water

---

* At Bald Mountain the gravel was brought from the mine a distance of over seven thousand feet, by a locomotive, drawing a train of eighteen cars, carrying two tons per car. The locomotive weighed seven and one half tons.   The trip was made in five minutes. Anthracite coal was used for fuel.   According to Superintendent Wallis, the relative costs of transportation at Bald Mountain were: By man power, 21 cents per carload (two tons); by mule, 9 cents per carload; by locomotive, 4¾ cents per carload.

tank, and the gravel is washed into and through the sluices. Where the water supply permits, the gravel is washed once a day. A permanent and sufficient water supply is very desirable for washing the gravel, but at many mines there is a scarcity of water, in consequence of which the gravel can only be washed at irregular intervals.* Sometimes the gravel is stored in bins for several months, awaiting available water; at some places the drain water from the mine is stored in a large water tank near the head of the sluices, and when a sufficient quantity has accumulated, is used for washing the gravel. Occasionally, water from this source is sufficient for this purpose, but usually the supply has to be supplemented by water from elsewhere. It requires from one sixth to one twelfth of an inch of water to wash a cubic yard of " free " gravel. The sluices are wooden boxes from eighteen to twenty-four inches in width and depth, and usually twelve feet in length. They are set on a grade of eight to eighteen inches per twelve feet, and paved usually with iron riffles that may be readily removed. The length of the string of sluices is rarely more than a few hundred feet—most of the gold is caught within one hundred feet of the head of the bin when the gravel is very free. Drops and undercurrents should be more frequently used in this connection than is the practice, and more quicksilver added to the sluices. At many mines, where the gravel is somewhat cemented, the tailings are impounded below the sluice and allowed to slack, after which they are rewashed by a second string of boxes below the dam.

The milling process is adopted when the conglomerate is too strongly cemented to be disintegrated (as in the sluicing process) by means of water. The gravel must then be crushed in a gravel mill, which, in most respects, is similar to the mills employed in the reduction of gold ores. The gravel is crushed through coarse screens, three sixteenths of an inch mesh in diameter, and is mostly caught in the battery by amalgamation. Outside copper plates, as in gold milling, are also used. The capacity of gravel mills is from five to twelve tons per stamp in twenty-four hours. It takes from one third to one fourth of an inch of water to mill a cubic yard.

Mills for gravel should have double discharge mortars. They should have rock breakers and self feeders. No attempt is made to save the sulphurets, which are usually found in the conglomerate. These sulphurets (and magnetic sands) are auriferous, but generally not rich or abundant enough to pay for saving. The cost of mining by this method is, of course, exceedingly variable.

COSTS OF DRIFT MINING.

Where the deposit is worked by tunnels and the gravel is free (not cemented), under favorable conditions gravel may be mined and washed for 90 cents to $1 25 per cubic yard; where the gravel is cemented and requires much blasting the cost of mining will be considerably increased. Such gravel also has to be milled, further enhancing the cost from 20 cents to 40 cents per ton. Under favorable conditions of mining and milling, cemented gravel may be worked at a cost of from $1 75 to $3 50 per cubic yard.

The mines are ventilated in the most improved systems by blowers (Baker's, etc.) run by machinery. This is a cheaper and far more effective system than ventilation by means of air shafts and air drifts kept open for that purpose. The latter system costs from $3 to $5 per lineal foot of channel worked, while the former costs less than 50 cents.

---

*The dumps where the gravel was stored awaiting water for washing, at the Bald Mountain Mine, had a capacity of twenty-six thousand and sixteen thousand carloads, respectively.

EXAMPLES OF DRIFT MINES.

May Flower.—Situated in Placer County. From December 11, 1888, to September 24, 1889, the company's twenty-stamp mill crushed thirty-three thousand seven hundred and eighty-seven tons of cemented gravel, which yielded $272,616 50, or 8\frac{6}{100}$ per ton. But few mines have as rich gravel as this. The monthly expenses of the company ranged from $10,000 to $13,000. A lineal distance on the channel of one thousand six hundred and twenty feet yielded the thirty-three thousand seven hundred and eighty-seven tons referred to.

Hidden Treasure.—Situated on the Damascus Channel (see sketches), Placer County. The gravel worked is not cemented, but is cheaply mined and is "free milling;" *i. e.*, it does not require crushing. This ground is "picking" ground, requiring no blasting. The timbering expense is, however, much greater than that of the cemented gravel mines. From February 27, 1888, to June 30, 1888—one hundred and eight working days—the figures are:

	Per Load— One Ton.	Total.
Gold yield	$1.2347	$39,821 53
Wages	$0.7202	$23,528 00
Contracts	.1077	3,464 78
Expense, material, etc	.0957	3,086 94
Total expense	$0.9236	$30,079 72
Profit	.3111	9,741 81

Number of days' labor, 11,164.50. Number carloads gravel, 32,252.

For the eleven years, 1877 to 1887, inclusive:

Receipts.

Gold yield	$879,523 27
Receipts from other sources	19,176 16
Total	$898,699 43

Expenditures.

Wages	$490,297 64
All other expenses	137,064 35
Dividends	268,092 00

The cost per carload ($0.9236) is exceptionally low, as under ordinarily favorable conditions $1 50 to $1 75 a carload is as low a figure as can be anticipated, and in most of the mines the cost is from $2 to $3.

The yield of this mine last year was, approximately, gross: $81,000; expenses, $51,000; net, $30,000. The value of the gravel per ton was $1 56. One hundred and twenty men were employed.

Bald Mountain Mine.—Situated in Forest City, Sierra County. The owners of this property invested $20,000 in opening up the mine, with the following gratifying results:

Year Commences July First—	Carloads.	Gross Yield.	Dividends.
1872–1873	50,168		
1873–1874	65,782	} $544,000 00	$284,000 00
1874–1875	79,990		
1875–1876	100,080	296,341 76	150,000 00
1876–1877	98,044	235,803 57	70,000 00
1877–1878	106,160	269,755 00	120,000 00
1878–1879	90,274	164,909 00	40,000 00
1879–1880	86,378	188,892 40	60,000 00
1879–Oregon Creek		3,000 00	
	676,876	$1,702,701 73	$724,000 00
July, 1880		16,914 38	20,000 00
July, 1880, 920 feet, Oregon Creek		9,000 00	
August 1, 1880		$1,728,616 11	$744,000 00
From June, 1881, to June, 1887 (mine closed down, having been worked out)		803,124 00	240,000 00
Totals		$2,531,740 11	$984,000 00

In addition to this, upwards of fifty thousand was taken from the reworked tailings which had been run into the cañon to slack (disintegrate the cemented gravel). From 1881 until 1887 (when operations were discontinued) the gravel output was from sixty thousand to one hundred and seven thousand carloads (one and one fourth tons per car) annually. The costs of mining, washing the gravel, and general expenses, etc., were from $1 22 to $1 44 per carload of extracted gravel. During 1881–82 the gravel yielded 50⅛ cents per square foot of ground removed. The Ruby Gravel Mine, on the same channel as the Bald Mountain, has a pay lead of from fifty to three hundred feet in width (including the benches of gravel). Under the economical management of Mr. Pichoir, gravel is extracted from the breasts twelve thousand feet from the mouth of the tunnel, and washed at a total cost of only $1 25 per carload of one and one fourth tons. This gravel is free, and requires no blasting.

The record for this channel to the year 1880 is as following—three thousand eight hundred and fifty feet worked:

Gross yield .... $1,788,000; $465 per lineal foot of channel.
Working expenses .... 923,000; $240 per lineal foot of channel.

Profit (including plant) .... $865,000; $225 per lineal foot of channel.

This company owned seven thousand five hundred feet of channel.* The claim was worked out about three years ago. In full operation the mine employed from one hundred and twenty-five to one hundred and seventy-five men.

HYDRAULIC MINING.

The recent suppression of hydraulic mining by judicial decisions, has reduced the annual gold product of California by at least $10,000,000; has thrown thousands of men out of profitable employment, and has withdrawn enormous sums of money from circulation in the various channels of trade. Nor have the hardships which resulted from this been confined to persons directly connected with mining enterprises; being felt generally throughout the community, and they have fallen to a great extent on the very persons whom the decisions in question were designed to protect. For-

---

* Lava flow carried away five hundred feet of this channel.

tunately, however, the interdependence of the various industries of the State in this respect is becoming generally recognized, and thoughtful men, of whatever profession, are awakening to the desirability of rehabilitating the hydraulic mining of the State. It is greatly to be hoped that some method will shortly be devised for the accomplishment of this end, and for the effective prosecution of hydraulic mining, *in a manner which will at the same time insure to the farming interests of the State the protection to which they are entitled, and preserve the navigable rivers of the State as well.*

In general, hydraulic mining consists in the disintegration of the auriferous alluvia, by propelling a heavy jet of water under pressure upon the bank, and in washing off the gravel in sluices in which is distributed mercury. The gold forms an amalgam and remains caught. The determining conditions of the profitable prosecution of hydraulic mining are:

First—Attainment of a large supply of water under a high pressure at not too great a cost.

Second—Facilities of obtaining the grade requisites for the sluices, and the dump for the refuse or "tailings."

MEASUREMENT OF WATER.

The method by which water is measured in the mining counties of California is that in vogue in Italy and Spain.

An aperture, whose sectional area represents a certain number of square inches, is cut through an upright or vertical board, which forms a portion of a confined box. The amount of water discharged through this orifice, divided by the number of square inches of the sectional area of the orifice, is called the "miner's inch."

The quantity of water represented by the "miner's inch" varies throughout the State by reason of the difference in pressure (*i. e.*, the height of the surface of water above the orifice), the thickness of the board through which the orifice is cut, etc., and ranges from two thousand to two thousand six hundred cubic feet per twenty-four hours of flow. The quantity of water discharging through an orifice one inch square, through a two-inch board, constitutes the "miner's inch" most generally adopted throughout the State. The quantity of water discharged under the above conditions for ten hours is called the ten hours miner's inch. The twenty-four hours miner's inch is that understood when not otherwise expressed. The miner's inch, as above described, is approximately sixteen thousand eight hundred gallons.

WATER SUPPLY.

The water used in hydraulicking is derived from the streams fed chiefly by the rains and melting snows. Where operations are conducted upon a comparatively small scale (the companies having no storage reservoir of any considerable capacity), the hydraulicking season is about coextensive with the rainy season, rarely being prolonged more than a month or two after the cessation of the rains. Under such conditions five to six months is about the duration of hydraulicking.

NOTE.—In some localities in California, Idaho, Montana, and Colorado, "booming" is practiced. This system is adopted where there is a scarcity of water, or where the gravel is too limited in extent, or too poor in grade, to justify the expense of bringing in a permanent supply of water for hydraulicking. It consists in accumulating a supply of water in a reservoir at the head of the gulch where the gravel is to be washed, and discharging (usually automatically) the entire volume of water at once, so that the rush of the water carries off the gravel into sluices disposed for that purpose. Some of the gold is caught in the sluices—a great part obviously is carried off. Booming is confined to the gulch gravel.

NORTH BLOOMFIELD MINE, NEVADA COUNTY.

BOWMAN (MAIN) DAM, NORTH BLOOMFIELD COMPANY.

Above sea level, 5,500 feet; height above bed of cañon, 98 feet; length on top, 420 feet; thickness at bottom, 167 feet; contains 55,000 cubic yards of stone; capacity of reservoir, 907,000,000 cubic feet, or 410,000 twenty-four-hour inches, or sufficient to supply San Francisco for ten years; water brought by ditch to mines, distant 45 miles; average yearly rainfall at dam, 65 inches.

BLUE TENT SYSTEM.
Flumes and ditches carried around steep and rocky mountain side.

Companies hydraulicking more extensively have storage reservoirs commensurate with the magnitude of their operations. Such companies are enabled to continue their piping with but little interruption during the entire year. The reservoirs constructed for mining purposes (chiefly for hydraulic mining) on Yuba, Bear, Feather, and American Rivers have an aggregating storage capacity of about fifty billion gallons, which is about twice as much as the Spring Valley Water Company's system.

The source of water supply òf the North Bloomfield Gravel Mining Company, the Milton Mining and Water Company, and the Eureka Lake and Yuba Canal Company (consolidated), is about the headwaters of Big Cañon Creek and Middle Yuba River in Nevada and Sierra Counties. The catchment area embraced in those sections represents in the aggregate 68.6 square miles. There are eleven principal reservoirs varying in area from ten acres to four hundred and eighty-seven and one half acres (high water area) each, and having a capacity of from $2\frac{1}{2}$ to 796.7 million cubic feet. The total area of the reservoirs of those companies is over eleven thousand six hundred acres (at high water mark) with a total capacity of over two billion one hundred and ninety-five million cubic feet.

In order to obtain efficient head or pressure, it is often necessary to bring the water from great distances. To overcome the topographical obstacles, considerable engineering skill is sometimes required. In Butte County the bracket flume of the Miocene Ditch Company is something unique in hydraulics.

In order to obviate the construction of a trestle some one hundred and eighty feet high, the water is conveyed in a wooden flume (four feet wide and three feet deep) around a bluff three hundred and fifty feet in height. The flume was suspended upon brackets made of T rails, bent in the form of a reversed L (⌐) soldered into holes previously drilled into a solid vertical escarpment; men were swung down by ropes to drill these holes.

In another place in this line of ditch is a piece of trestlework one thousand and eighty-eight feet long and eighty feet high.

Herewith is given the statistics of the water companies in the central mining counties of the State. The water of these companies is used principally for mining purposes.

Water for hydraulicking costs 5 to 25 cents an inch, when purchased from water companies; 10 to 15 cents is the usual price paid by hydraulic mining companies.

In 1877–78 the Bloomfield Company used eight hundred and fifty-five thousand miner's inches (twenty-four hours inches) of water, at a cost of $2\frac{6}{10}$ cents per inch.

La Grange Ditch and Hydraulic Mining Company, Stanislaus County. Water from Tuolumne River, eighteen miles from mine. Length of ditches, twenty-five miles. Dimensions: top, nine feet; bottom, six feet wide; depth, four feet; capacity, two thousand seven hundred miner's inches per twenty-four hours; grade of ditches, seven to eight feet per mile. Cost of ditches, etc., $450,000.

Tuolumne County Water Company, Tuolumne County. Water from south fork of Stanislaus and tributaries. Miles of ditches, one hundred and twenty-five (fifty miles not in use now); six miles of flumes. Grade, ditches, eleven to thirty-two feet per mile. Dimensions of ditches: width, bottom, seven and a half to eleven feet; top, eleven to fifteen feet; depth, four feet. Ditches cost, on average, $4 50 per yard; flumes, $14 per yard; and pipes, $6 per yard.

Union Water Company, Calaveras County. Water from north fork of Stanislaus River. Forty miles of ditches; capacity, two thousand five hundred miner's inches; water grade, three to twenty-five feet per mile. Total cost of plant, $200,000.

Mokelumne Hill and Campo Seco Ditch Company. Water from head-waters of Mokelumne River. Ditches, over one hundred miles; grade, eight to sixteen feet per mile; capacity of ditch, one thousand five hundred inches. Cost, about $500,000.

El Dorado Water and Deep Gravel Mining Company, El Dorado County. Main reservoir, Silver Lake, Amador County. Main ditch, forty miles; tributary ditches, seventy miles; total ditches, one hundred and ten miles; one mile of flumes; three miles of pipes. Dimensions of ditches: top, ten feet; bottom, six feet; depth, four feet; grade of ditches, four feet to mile; flumes, grade, one and one third to two and one half feet per mile; water delivered, four thousand inches. Reservoirs cost $45,000. Total cost of plant, $600,000.

California Water and Mining Company. Water from Loon Lake and Pilot Creek, in El Dorado County; two hundred and fifty miles of ditches; grade, six to sixteen feet per mile. Dimensions: top, three and one half to eight feet wide; bottom, two to five feet wide; one and one half to three feet deep; water supplied by ditches, one thousand two hundred inches; flumes, two and one half miles. Total cost of plant, $600,000.

Park Canal and Mining Company. Water from different branches of Cosumnes River; ditches, two hundred and ninety miles; flumes, eight miles; pipe, one mile; grade of ditch, one half to sixteen feet per mile. Dimensions: top, eight feet wide; bottom, five feet wide; two and one half feet deep; capacity of ditches, two thousand two hundred inches. Reservoirs cost about $60,000. Total cost of plant, $2,000,000. Ditches cost $10 per rod; flumes, $12 to $14 per rod; pipes, $2 50 to $4 50 per yard.

Iowa Hill Ditch Company, Placer County. Water from North Fork of American; twenty-five miles of ditches; capacity of ditch, three thousand inches. Plant cost $200,000.

North Bloomfield Company, Nevada County. Length (including reservoirs) of ditches, one hundred and fifty-seven miles; capacity, three thousand two hundred inches; grade, twelve to sixteen feet per mile. Dimensions of ditch: top, eight and two thirds feet wide; bottom, five feet wide; depth, three and one half feet. Cost of plant, $708,841. It costs $13,463 per year to keep the reservoirs and ditches in order.

Milton Company (including reservoirs). Length, eighty miles; grade per mile, twelve to twenty-five feet. Dimensions: top, six feet wide; bottom, four feet wide; depth, three and one half feet; capacity, three thousand inches. Cost, $391,579.

Auburn and Bear River Canal Company, Placer County. Seventy-five miles of ditches; capacity, three thousand inches. Cost, $350,000.

Amador Canal, Amador County. Ditches, sixty-six miles; capacity, two thousand inches. Cost, $400,000.

Brandy City. Ditches, seventeen miles; capacity, two thousand inches. Cost, $150,000.

Buckeye Company, Trinity County. Ditches, thirty-five miles; capacity, two thousand five hundred inches. Cost, $120,000.

Dardanelles Ditches, Placer County. Seventeen miles; capacity, three thousand inches. Cost, $125,000.

Del Norte Company. Ditches, ten miles; capacity, two thousand inches. Cost, $40,000.

Gold Run Ditch Mining Company. Ditches, twenty-six miles; capacity, two thousand five hundred inches. Cost, $150,000.

SPRING VALLEY MINE, BUTTE COUNTY.

Two streams piping. Gravel is washed into the deep bedrock cut which leads to the tunnel seen in background, right hand side of picture. This tunnel is three thousand feet long, and was driven to get outlet for tailings.

BRITTON & REY, S. F.

HOBSON MINE, PLACER COUNTY.

Monitor playing five hundred miner's inches under head of three hundred and sixty feet. It will be seen that the bottom gravel is much coarser than the gravel higher in the bank.

Little York and Liberty System, Nevada County, Cal. Ditches, thirty-five miles; capacity, three thousand five hundred inches. Cost, $150,000.

Natoma Water and Mining Company. Ditches, sixteen miles; capacity, three thousand five hundred inches. Cost, $390,000.

Phœnix Ditch Company. Ditches, one hundred miles; capacity, four thousand inches. Cost, $880,000.

Powers' Ditch, Butte County. Ditches, thirty miles; capacity, two thousand inches. Cost, $75,000.

California Water Company. One hundred and twenty-five miles; capacity, four thousand five hundred inches. Cost, $550,000.

Eureka Lake and Yuba Ditch Company. Length of ditches, one hundred and sixty-three miles; capacity, five thousand eight hundred inches. Cost $723,342.

South Yuba Ditch Company. One hundred and twenty-three miles of ditches; capacity, seven thousand inches. Total cost of plant, $1,100,000.

Smartsville Ditches. Capacity, five thousand inches; grade, nine feet per mile; cost, $1,000,000. Dimensions: top, eight feet wide; bottom, five feet wide; depth, four feet.

Spring Valley and Cherokee. Ditches, length, fifty-two miles; capacity, two thousand two hundred inches. Cost, about $500,000.

Hendricks. Ditches, length, forty-six and one half miles; grade, six to twelve feet per mile. Cost, $136,150.

Blue Tent. Ditches, thirty-one miles; capacity, two thousand one hundred inches. Cost, $200,000.

When feasible, ditches are used as conduits in preference to flumes or pipes, as the cost of construction and of maintenance is less than that of the flumes or pipes.

In some places, the topography of the country, the character of the ground (hardness, porosity, etc.), or other conditions, render the use of flumes more economical. Flumes are usually of smaller sectional area than the ditches, but are given more grade (twenty to forty feet per mile) to compensate for the reduced area. Flumes cost from $1 to $2 per linear foot.

At Cherokee, in Butte County, the water is conveyed across a deep ravine by an inverted siphon of wrought iron. The diameter of the pipe is thirty to thirty-four inches, and its greatest thickness, where subjected to a pressure of eight hundred and eighty-seven feet (three hundred and eighty-four pounds per square inch), was No. 00 iron, Birmingham gauge, 0.375 inches.

DUTY OF MINER'S INCH.

The duty of a miner's inch of water is the quantity of material which that amount of water is capable of moving. Obviously, the duty of a miner's inch will vary greatly, depending as it does on the quantity of water used, the pressure of the water, the character of the material washed, and the grade and width of the sluices.

NAME OF STREAM.	Quantity of Water Used in Mining and Discharged into Beds of Rivers in Twenty-four Hours—Inches.	State Engineer's Estimate of the Duty per Inch—Cubic Yards.	Amount Moved—Cubic Yards.
Table Mountain or Dry Creek	833,250	3½	2,916,375
Butte Creek	24,000	3	84,000
Feather River	1,259,363	3½	4,407,770
Yuba River	5,458,171	3½	19,103,598
Bear River	1,117,082	3	3,351,246
Dry Creek, No. 2	44,229	3	132,687
American River	1,914,500	4½	8,615,250
Totals	10,650,595	* 3.6	38,610,926

* Average.

These estimates are less than the actual results. Other conditions being the same, the duty of a miner's inch increases rapidly with an increased grade of the sluices. Le Conte says the transporting power of water is as the sixth power of its velocity.

At Hobson's mine, in Placer County, piping a bank one hundred and twenty feet high, of very light free gravel containing no cement, using five hundred inches of water, under a pressure of three hundred and sixty feet, with a twelve-inch grade (twelve inches to a box of twelve feet) for sluices, the duty of a miner's inch was twenty-four cubic yards. At the same mine, with the same grade, same quantity and pressure of water, but where the gravel was coarse and cemented, the duty was reduced to ten yards.

Under the same conditions of gravel bank and water supply as referred to in the first example given, but where the grade of the sluice was increased to eighteen inches a box, and where iron riffles were used instead of rocks, the duty of thirty-six cubic yards per inch was attained. These results are exceptionally high in hydraulicking. At Wisconsin Hill, Placer County, piping light top gravel with a grade of twelve inches per box of twelve feet, with five hundred inches stream, the duty of a miner's inch was ten cubic yards, while only three cubic yards of bottom and cemented gravel were washed under the same conditions.

At the North Bloomfield, in washing one million five hundred thousand cubic yards of top gravel, the duty of a miner's inch was 5.39 cubic yards.* In washing seven hundred and ten thousand cubic yards of the underlying gravel (to height of sixty-five feet above the bedrock) the duty of an inch was but 2.34 cubic yards. In the latter case, thirty-five cubic feet of water were required to move one cubic foot of gravel.

DUMP, ETC.

Less indispensable even than a good water supply is the available fall upon the property to secure sufficient grade for sluicing and satisfactory dumping ground for the tailings. Much room is required below the mine upon which to deposit the debris, where operations are conducted on a large scale. Where upwards of two millions of cubic yards of material

* At Bloomfield, twenty-four hours miner's inch equals two thousand two hundred and thirty cubic feet, the duty per miner's inch was ten cubic yards, while under the same conditions of pressure and grade of sluices, piping the underlying heavier cemented bottom gravel, the duty was but four cubic yards.

WATER POWER DERRICK FOR REMOVING HEAVY BOWLDERS.

MONITOR.

are moved annually, the mine, obviously, must have an extensive outlet for the debris. Deep cañons are the most favorable sites for this purpose.

In order to utilize the pressure due to the elevated position in which the water is brought, with respect to the gravel to be washed, the water is conducted from the ditches into a tank called the "pressure box" or "bulkhead." From the pressure box or bulkhead, by means of a feed-pipe (main pipe), the water is brought to a distributer. The size of the feed-pipe is determined by the quantity of water to be used. Twenty-two-inch (in diameter) mains are generally used in the larger hydraulic workings.

The pipes are made of wrought-iron, the thickness of which increases with the diameter of pipe, and the hydrostatic pressure to which the pipe is subjected, Nos. 16, 14, 12 (Birmingham gauge). These numbers correspond to thickness of .065, .083, 1.09 inches, respectively. To prevent the pipes from corroding, they are coated with a preparation of asphalt and coal tar. The pipes are made in lengths of twenty feet, and jointed together in stove-pipe fashion, rivets being rarely used.

The "distributer"* is a cast-iron box which serves the purpose of a hydrant, and by means of valves enables the partition of the stream of water and the diversion of the branch streams into two or more pipes, whereby more than one part of the gravel bank can be simultaneously hydraulicked.

From the distributer the streams are piped to the "monitors" or "giants." These are the discharge pipes which concentrate the stream and enable its projection upon any desirable point. The giants and monitors are labor saving. Before their introduction at North Bloomfield, there were ten to fifteen streams playing, while now one stream, tended by one man, does as much work.

The "nozzles" of the monitor are from four to nine inches in diameter. Two or more monitors are employed, depending upon the magnitude of the workings. The streams played upon the bank are one hundred or more feet in length. The large mines have a dozen or more monitors, but rarely have enough water to supply at once more than four or five monitors.

The disintegrating power of the jet, which at the larger mines equals one thousand to one thousand five hundred inches of water (one thousand five hundred to one thousand seven hundred and fifty cubic feet per minute), weighing sometimes upwards of one hundred thousand pounds, under pressure of one hundred and fifty feet to four hundred and fifty feet, is enormous. Yet, notwithstanding this, the gravel is sometimes so tenaciously cemented that assistance of powder becomes necessary to shatter or break up the bank preparatory to its disintegration by water.

It is desirable to get the monitor or giant as near the bank as is consistent with the safety of the miners and the machine, in order to obtain the full force of the stream. Where the banks are high—over two hundred feet or so—they are accordingly usually worked in benches of from one hundred and fifty feet to two hundred feet in depth.

The system of "bank blasting" generally in vogue is as follows: A small level, or drift, is run into the bank at right angles to the face of the bank, and from the end of this drift crosscuts are driven parallel to the bank (at right angles to the drift). The drifts will form a T-shaped excavation. The length of the main drift will vary from twenty to one hundred feet, and in the cross-drifts, of which there may be two (forming a ⊤), the kegs of powder are placed. The cubical contents of the portion of the bank to

---

* Owing to the liability to break, and their unwieldiness, the cast-iron distributers have, in many places, been displaced by sheet-iron branch pipes, or tees, which have cast-iron gates.

be blasted, the tenacity of the material of the bank, etc., will determine the length of the powder drifts and the quantity of powder to be used. From a few dozen to a thousand or more twenty-five-pound kegs of black blasting powder * are ignited in one blast. The drifts are well tamped, and the charges are simultaneously fired by means, generally, of a high-tension electrical machine.

Where the banks are worked in "benches" or stopes, vertical shafts ten to twenty feet deep are sunk from the surface of the deposit, short drifts are run from the bottom of the shaft, forming an inverted T ($\perp$), and the powder charged, tamped, and fired as above. Very much lighter charges are used in this kind of blasting. A slow lifting powder is generally used for bank blasting, while the nitro-glycerine explosives are used for blasting the bowlders, trees, etc., which were washed down into the pit or open space on the bedrock left after washing away the gravel, etc.

The second condition prerequisite to the successful working of the hydraulic mine is the attainment of the grade necessary for the treatment of the detrital material piped from the bank.

To get the requisite grade for the sluices, and, at the same time, to obtain a suitable place for the deposition of the debris from the washing out of the gravel bank, often involves the driving of a long bedrock tunnel.

The topographical features of the environs will determine the location of the tunnel. The mouth of the tunnel should be sufficiently below the bedrock of the deposit to be prepared for the contingency of a change in the grade of channel, whereby the tunnel might be placed above the level of the drainage and rendered practically useless. This difference of level also admits of the use of chimneys or shafts, connecting the surface of the bedrock of the channel with the face (interior end) of the tunnel, whereby a drop is obtained which facilitates the disintegration of the obdurate conglomerate. A line of sluices is laid in the tunnel.

The tunnels are from a few hundred to several thousand feet in length. The tunnel of the North Bloomfield Company, in Nevada County, is seven thousand eight hundred and seventy-four feet long, and its dimensions seven by eight feet. It cost about $500,000. The tunnels are from four to eight feet wide and from five to nine feet high.

The gradient to be given the tunnel will be determined by the fall available, the character of the material to be washed, etc. Tenacious material, such as very compact conglomerate, very indurated pipe-clay, etc., will require a high grade, and from ten to twelve inches per twelve feet would be advantageous.

The material "piped" from the bank is carried by the stream of water through bedrock cuts and sluices to the chimneys or shafts (where such exist).

The bedrock cut, as its name implies, is a trench carried in the bedrock from the upper end of the line of sluices to the gravel bank. These cuts are about as wide as the sluices, and are sometimes twenty to thirty feet deep. The gravel is washed into these cuts and thus is brought to the sluices. These cuts are not paved. Cuts require nearly twice as much grade as sluices.

SLUICES.

These are commonly called "flumes" in mining parlance, but the term "flume" should be restricted in meaning to a conduit for carrying water to be used in hydraulicking, while the term "sluice" in contra distinction

---

* Low-grade dynamite powder has almost entirely superseded black powder in bank blasting.

SLUICE LINE CONSTRUCTED ON CURVE.

should refer to the boxes or troughs below the cuts through which the gravel is washed, and in which the gold is recovered.

These are a kind of water trough or box from three to six feet wide, and from two to three feet deep, and most generally twelve feet in length per box. From one hundred to several hundred boxes are used in a line of sluices.

PAVEMENTS.

To prevent excessive wear and tear of the sluices, they are lined with heavy planks on the sides and are paved with rocks and blocks which also serve as riffles to arrest the flow of the gold and amalgam. Iron rails ("T" rails from railroads), or wooden rails covered with bar iron, are placed longitudinally upon the bottom of the sluices and also serve as riffles.

Economic conditions will determine which of the styles of riffles are used. Rock riffles will wear longer than the other kind, lasting from three to six months, but require more grade to the sluices, and also a loss of more time in cleaning up and repaving the sluices.

Block riffles are made, where possible, of the "digger" pine (*Pinus sabiniana*) and other pine. Hard wood is not as good as the softer pine, which has the property of brooming up and thus presenting a better surface to arrest the gold and amalgam. They are square, varying in size from twelve to thirty inches, and in depth from ten to eighteen inches. The interstices of the pavement are filled with small stones. Block riffles last from two to four weeks in the average conditions of hydraulicking. Iron riffles considerably longer. Though more costly in the first instance, they are cheaper in the end, owing to their longer life and greater economy of time in being changed.

GRADE OF SLUICES.

Where top and poor gravel is being piped it is desirable to run it off as fast as possible. The grade upon which the sluices are set will be regulated by the available fall along the line of sluices, and by the character of the material washed. In some localities the adoption of minimum grades two to four inches per box is enforced by lack of fall. Not only does the use of such light grades greatly decrease the duty of water, but it involves a large increase in the expense of handling the rocks not capable of being sluiced upon such grades. The grade in general use is what is known as a six-inch grade (six inches to a box of twelve feet), but where practicable the use of grades from eight to twelve inches are more advantageous. Steep grades facilitate the thorough disintegration of the cemented gravel, and thereby effect a reduction of the length of sluice line, which would otherwise be necessary to insure this result.

It is best to have steep grades for the disintegration of the gravel and depend upon the undercurrents to save such gold as would not be deposited on account of the grades of this character. Steep grades are especially advisable where water is expensive or scarce. These sluices are several hundred and sometimes several thousand feet in length. (See "Loss of Gold.")

Some mines have a double line of sluices, so as to avoid loss of time in cleaning up and repaving sluices.

Sluices cost from $25 to $35 per box (twelve feet) at the larger mines. The Spring Valley Mine has three parallel lines of sluices, two and a half miles in length each.

GRIZZLIES AND UNDERCURRENTS.

When the available fall admits of it, one or more grizzlies and undercurrents are used along the line of the sluices.

A grizzly is a grating or framework of iron bars, laid parallel, with interstices between to allow the finer material to fall into the sluices, or undercurrents, below, while the coarser barren bowlders are screened off, as it were, and dumped outside of the flume.

It is desirable to get rid of the large bowlders as soon as possible, to prevent unnecessary wearing of the pavement of the sluices and waste of water in washing them; but where the gravel is much cemented the bowlders assist in its disintegration.

To remove the large bowlders (sometimes weighing several tons) from the bedrock and cuts, a kind of derrick crane is used. These bedrock derricks have a mast from eighty to one hundred feet high and a boom eighty to ninety feet long, and are modified in other respects to best adapt them to their employment in hydraulic mines. The actuating power of the derrick is, generally, a hurdy-gurdy. This is a peculiar kind of impact wheel, made to utilize water under high pressure. The water is projected through a nozzle tangentially to the wheel, upon the periphery of which are set radial buckets. The Pelton and Knight's wheels are examples. Under favorable conditions, the Pelton wheel will develop about eighty per cent of the theoretical power of the water.

The undercurrents are large boxes or tables, of various shapes and sizes, more commonly ten to twenty feet wide and forty to fifty feet long, which by the distribution of the water over larger surfaces makes the stream shallow and allows the deposition of the gold and amalgam upon the riffles with which they are paved. Block, rock, or iron riffles are used in the undercurrents. The fine material which falls through the interstices of the grizzly is usually carried to the undercurrents.

AMALGAMATION.

Quicksilver is added several times a day, in quantities depending upon the length of the sluices. The more quicksilver added the greater are the chances of catching the gold. The usual practice is to sprinkle the quicksilver in the sluices and undercurrents, the greater part being added near the head of the sluices. No quicksilver is added to the cuts. From the consistency of the amalgam, and the appearance of the quicksilver in the sluices (which are frequently inspected), the feeding of the quicksilver is regulated.

Two or four tons of quicksilver are often in process of manipulation at once at some of the large hydraulic mines, one part of it being in the sluices and another part on hand as amalgam. The upper portions of the sluices are cleaned up, generally once or twice a month, the middle portion less frequently, and the lower part only once a season.

With large "heads"* it is advisable to clean up frequently to prevent a loss of amalgam through prolonged trituration. The upper part of the sluices for a distance of about two thousand feet ought to be cleaned up about every two or three weeks, where large quantities of gravel are washed. A gang of miners can clean up and repave a thousand feet of sluices in a day and night.

---

* The term "head" is often used in this connection to express the quantity of water used, as, for example, a head of five hundred inches, etc. It must not be confounded with the hydrostatic head, which the miners commonly designate as "pressure."

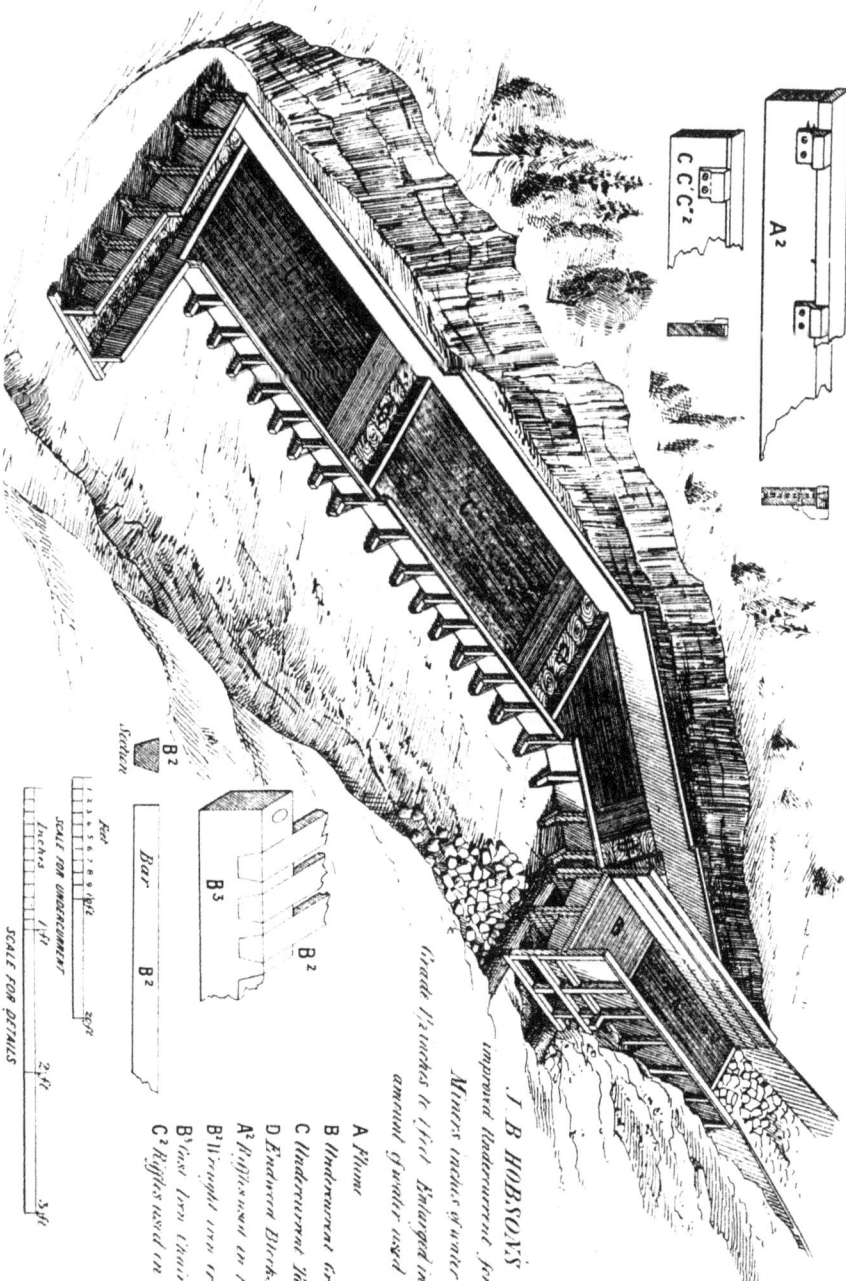

J. B. HOBSON'S

improved Undercurrent for 600

Miners inches of water.

Grade 1½ inches to 1 feet. Enlarged in proportion to
amount of water used.

A Flume

B Undercurrent Grating

C Undercurrent Tables

D Enlarged Blocks 6" thick

A² Riffles used in Flume

B¹ Wrought iron or Steel bars

B³ Cast Iron Chairs

C² Riffles used in Tables

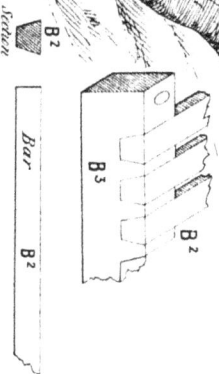

B²

Section

Bar

B²   B²

B³

B²

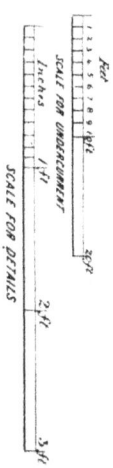

Feet

SCALE FOR UNDERCURRENT

Inches

SCALE FOR DETAILS

C C¹ C² ²

A²

DROP FOR DISINTEGRATING GRAVEL, AND UNDERCURRENT.
The material passing over the drop and that discharging from the undercurrent is taken up by sluices below.

METHOD OF CARRYING FLUME ACROSS CAÑON BY IRON BRIDGE.
Also grizzly for removing large bowlders. Lack of room prevents introduction of undercurrent at this point.

CHEROKEE FLAT, BUTTE COUNTY.

A nest of undercurrents. This company had twenty-four undercurrents along its line of sluices. Between the lowest two undercurrents is seen the grizzly over which the coarse bowlders are washed, while the fine material which passes through the interstices of the grizzly is carried over the under-

CLEAN-UP.

In cleaning up the riffles are removed, a small stream of water is run over the sluices, and the gold and amalgam collected by scoops, etc. The amalgam is strained, cleaned, retorted, and melted. With the amalgam many of the minerals described before as existing in the gravels are found.

From 75 to 80 per cent of the total yield of amalgam is obtained in the first three hundred to four hundred feet of sluices. A small percentage of the gold, in the form of dust, comes from the bedrock cuts. The remainder of the gold recovered comes from the undercurrents and in the lower section of the sluices.

Total yield for the year 1877-78	$311,276 20
Near bank, from rock cuts in mine (dust)*	4.57 per cent.
Flume in tunnel (1,800 feet)	86.96 per cent.
Tunnel below flume (6,000 feet)	4.50 per cent.
Cut below tunnel (200 feet)	0.81 per cent.
Tail sluices (300 feet)	1.21 per cent.
From seven undercurrents	2.65 per cent.
	100.00 per cent.

The first undercurrent caught five times as much as the sixth, and nearly three times as much as the seventh undercurrent, which was of double size. This last yield, $947, induced the company to add another undercurrent.

The smaller mines use cup-shaped retorts, as used in the smaller gold mills, while the larger mines use retorts, such as are used in silver mills. Bullion from the hydraulic mines is notably very much finer than that from the quartz mines. This bullion will run from 850 to 980 fine. The bullion from the Australian gold gravels is much finer than that of California.

Silver is generally the debasing alloy, though where the grade of bullion is very low lead or copper are often found in the alloy. The value of the amalgam in the upper portions of the sluices is from $7 to $12 per ounce. The amalgam from the lower part of the sluices, from the undercurrents, is of less value, on account of the fineness (in texture) of the gold caught there. The amalgam from top gravel is poor for the same reason.

QUICKSILVER LOSS.

The loss of quicksilver is, under the average conditions, from 10 to 15 per cent of the quicksilver used; sometimes the loss is as great as 30 per cent, especially where the gravel is much cemented.

At the larger mines piping goes on night and day. Pitch bonfires are generally used to illuminate the works during the night, but at the North Bloomfield an electric light of twelve thousand candle power was used, and was found more effective and economical. At Cherokee two electrical lights, of eight thousand candle power each, are used.

GOLD LOSS.

The loss of gold in hydraulic mining varies according to the character of the gravel washed and the system of sluices, undercurrents, etc., in use. Where the gravel is a hard, tenacious, conglomerate some gold is liable to be carried away because of the imperfect disintegration of the gravel; also the presence of pipe-clay increases the gold loss. Where such gravel

---

* No quicksilver is added in the cuts.

occurs it is desirable to disintegrate the gravel by blasting and by introducing, where practicable, numerous drops along the line of sluices. A long line of sluices with frequent drops and many undercurrents reduces correspondingly the gold loss. Of course, an appreciable amount of gold must inevitably escape as flour gold, rusty gold, and amalgam. Properly constructed undercurrents will diminish this loss, but still the last undercurrent of a series of many distributed over a long line of sluices (many miles in length) would undoubtedly catch some gold.

Other conditions not preventing, the length of the sluice system adopted is determined by the cost of the construction and of the maintenance compared with the value of the gold saved by reason of the increased length of the system.

No attempt has been made to introduce a system such as is employed in our best gold mills (see article on Gold Mills, Report State Mineralogist, 1888), to sample the tailings, and, therefore, it is difficult to approximate the percentage loss in hydraulicking operations. The examinations of many "tailings cañons," into which the debris from the gravel mines was washed, lead me to believe that but a small percentage of the gold is lost. It is often asserted that not more than one half of the gold in the gravel bank is saved by this method of mining. In the opinion of the writer, in most well conducted dydraulic and drift mining operations, at least 85 per cent and in many cases upwards of 95 per cent of the gold tenure of the gravel is saved. Mr. Louis Glass, formerly the manager of the Spring Valley Mines, is of the opinion [that not more than five per cent of the gold passed off in the tailings.

TAILINGS.

By far the greater portion of the material washed from the banks finds lodgment within a short distance of the tailings' dump of the mine. Fortunately the cañons into which most of the mines tail (*i. e.*, wash the debris) are of no value for purposes other than to serve as storage reservoirs for the mining debris. A properly constructed brush or log dam, preferably the former, for which most of the cañons below the mines afford advantageous sites, would, undoubtedly, impound nearly all the material which at present may be carried to the subjacent farming lands. Coöperative action in this matter, by companies having a common outlet for their tailings, would, in many localities, make it possible to operate their mines without damage to the agricultural or other interests. As these impounding dams become filled by the accumulated debris, they must be raised higher, or, if necessary, other dams must be constructed. But an insignificant amount of material would be carried over the dams as "slickens," in suspension into the valleys into which the cañons debouch. The cost of the construction and maintenance of a system of dams and of the canals for disposing of the "slickens" would be of no moment, as compared with the beneficial result of hydraulic mining.

EXTENT OF CLAIMS.

On account of the great expense usually attending the opening up of drift and hydraulic properties, and the low grade of the material to be mined, enterprises should be conducted upon a large scale. The larger drift or hydraulic mining companies own or control from one to five miles upon

NOTE.—A bar of bullion from the North Bloomfield mine weighed five hundred and ten pounds and was worth $114,290. This is the largest gold bar ever cast in this country.

BRUSH DAM, NORTH BLOOMFIELD COMPANY.

LOG DAM

For impounding tailings from an hydraulic mine. The material behind the dam is tailings from an hydraulic mine.

the supposed course of the channel. Some have invested in the property and plant from $1,000,000 to $3,000,000.

LABOR AT MINE.

The larger drift and hydraulic mines work from seventy-five to one hundred and twenty-five men, with wages ranging from $2 to $3 50 per day.

YIELD OF GRAVEL AT HYDRAULIC MINES.

The value of gravel washed at hydraulic mines is estimated usually upon the basis of yield per cubic yard, per miner's inch of water used, or sometimes by the product per acre. A cubic yard of gravel is from one and one half to one and three fourths tons.

Resume of work done by the La Grange Company on all its claims, June 1, 1874, to September 30, 1876:

1,533,728 inches (2,159 cubic feet each) washed 2,275,967 cubic yards of gravel, which yielded 12,026.84 ounces. Troy = $231,893.

Disbursements.

	Total.	Per Cubic Yard.	Per Ounce Metal Produced.
Water	$17,307 62	$0.008	$1 43
Labor	82,345 70	.036	6 85
Material	21,788 35	.010	1 81
Official	11,244 94	} .006 {	94
Contingent	3,125 80		26
Taxes	1,130 41		09
Totals	$136,942 82	$0.060	$11 38

Average value of the ounce of metal (gold and silver) produced................$19 29
Average yield per cubic yard of gravel................................. .1019
Average amount of gravel washed per inch, cubic yards..................... 1.48

NOTE.—The reader is referred to the excellent treatise of Mr. A. J. Bowie, Jr., for additional data.

Yield of Gravel at Important Hydraulic Claims in California according to Verified Reports.

NAME OF CLAIM.	Location.	Cubic Yards Washed.	Gross Yield.	Yield per Cubic Yard.	Height of Banks in Feet.	Report of—	Remarks.
American Co.	Sebastopol, Nevada Co.	5,171,834	$1,241,240 30	$0 24	120	H. Smith, Jr.	
No. 8—1870-74	North Bloomfield, Nevada Co.	3,250,000	94,250 00	2.9	150	H. Smith, Jr.	Paid a profit of $2,232 84.
No. 8—1874-75	North Bloomfield, Nevada Co.	1,858,000	74,271 77	3.9	180	H. C. Perkins	The North Bloomfield
No. 8—1875-76	North Bloomfield, Nevada Co.	2,919,700	192,735 73	6.6	200	H. C. Perkins	produced, from 1867 to 1880,
No. 8—1876-77	North Bloomfield, Nevada Co.	2,293,930	290,775 42	12.7	265	H. C. Perkins	$1,455,680 47; the mining profits
North Bloomfield	North Bloomfield, Nevada Co.	30,000,000	2,610,000 00	8.7	150-350	H. C. Perkins	were $411,589 27. Hammond.
French Corral	French Corral, Nevada Co.	4,200,000	1,745,500 00	41.5	20-100	H. C. Perkins	The greater part of the top gravel had been removed previously.
Manzanita	Sweetland, Nevada Co.	5,780,000	1,489,000 00	26	50-150	H. C. Perkins	About one third of the top gravel had been removed previously.
McCarty's	Columbia Hill, Nevada Co.	3,000,000	345,643 10	4.3		J. D. Hague	
Sicard	Patricksville, Stanislaus Co.	155,347	20,197 07	13	38	J. Messerer	Top gravel.
Delaney	Patricksville, Stanislaus Co.					J. L. Jernegan	
Chesnau	Patricksville, Stanislaus Co.	27,250	11,000 00	40.4	18	A. J. Bowie, Jr.	
Chesnau	Patricksville, Stanislaus Co.	71,810	9,847 48	13	55	J. Messerer	
Chesnau	Patricksville, Stanislaus Co.	284,932	47,781 73	16	12-62	J. Messerer	Aggregate of 7 surveys checked by 1 survey, June, '74, to Oct., '76.
Chesnau	Patricksville, Stanislaus Co.	338,880	62,980 37	18.6	90	A. J. Bowie, Jr.	Includes the last.
New Light	Patricksville, Stanislaus Co.	667,347	45,511 81	6.8	35	A. J. Bowie, Jr.	Drifted previously in places.
New Light	Patricksville, Stanislaus Co.	683,244	45,444 65	6.6	24-60	J. Messerer	Aggregate of 5 surveys checked by 2 surveys.
Johnson	Patricksville, Stanislaus Co.	196,632	9,148 27	4.6	30	A. J. Bowie, Jr.	
New	Patricksville, Stanislaus Co.	17,796	773 72	4.3	42	A. J. Bowie, Jr.	
Kelley	La Grange, Stanislaus Co.	88,660	3,406 33	4	85	A. J. Bowie, Jr.	Result obtained from cleaning out a deep hole.
Kelley	La Grange, Stanislaus Co.	351,152	43,153 26	12.3	75	A. J. Bowie, Jr.	Previously drifted; heavy blasting; no profit.
Kelley	La Grange, Stanislaus Co.	701,685	15,770 34	2.2	100	A. J. Bowie, Jr.	
New Kelley	La Grange, Stanislaus Co.	161,032	8,852 31	5.5	40	J. L. Jernegan	Upper bench gravel.
New Kelley	La Grange, Stanislaus Co.	252,614	35,012 33	13.8	65	J. L. Jernegan	Top and bottom gravel.
New Kelley	La Grange, Stanislaus Co.	1,000,000	64,550 27	6.4	40-45	J. L. Jernegan	Includes the two last data.
French Hill	La Grange, Stanislaus Co.	252,611	35,136 72	13.8	45	J. L. Jernegan	Winter of 1876-77.
French Hill	La Grange, Stanislaus Co.	674,968	90,186 19	13.3	*10-48	J. Messerer	Aggregate of 5 surveys checked by 2 surveys, May, '74, to Oct., '76.

French Hill	La Grange, Stanislaus Co.	1,020,317	188,433 11	15.5	30	A. J Bowie, Jr.	Includes the last and also early workings, of which portions had been previously drifted.
Light	La Grange, Stanislaus Co.	746,640	64,714 27	8.6	48	A. J. Bowie, Jr.	Banks contained several thick strata of sand.
Blue Point	Smartsville, Yuba Co.	93,944	115,728 17	1 23	57	H. Smith, Jr.	
Green Flat	Plumas Co.	22,000	15,000 60	67.5	15	A. J. Bowie, Jr.	
Fale's Hill	Plumas Co.	25,000	4,794 49	19	75	A. J. Bowie, Jr.	
Crawford's	El Dorado Co.	77,880	35,046 00	45	85	J. J. Crawford	
Gold Run District	Placer Co.	43,000,000	2,074,356 00	4.8	?	W. H Pettee	

* Average thirty feet.

Yield of Gravel, including Smaller Hydraulic, Drift, and Cement Claims, according to Authorities Given.

Name of Claim.	Location.	Cubic Yards Washed.	Gross Yield.	Yield per Cubic Yard.	Height of Banks in Feet.	Authority.	Remarks.
Bennet	Calaveras County	963	$1,320 00	$1 37	13	J. Rathget	Calculated from data in Raymond's Rep, 1872.
Johnston	Calaveras County	2,258	1,540 00	68.5	12½	J. Rathget	
Hedwick's	Calaveras County	2,963	1,450 00	48.5	20½	J. Rathget	
Whiteside's	El Dorado County	9,722	100,000 00	10 28	70	Raymond's Rep, '75	Calculated from data, p. 84.
Nagler	El Dorado County	20,000	100,000 00	5 29	30	Raymond's Rep, '75	
Spanish	El Dorado County	1,422	13,000 00	5 00	4	Raymond's Rep, '74	
Blue Tent	Nevada County	5,138,150	780,000 00	15			Estimated by several engineers, p.19.
South Yuba (G. Hill)	Nevada County	632,533	79,099 15	12.6		Official Rec., 1876-77	See "Contributions to Am. Geology," Vol. I, Whitney, p. 414.
South Yuba (G. Hill)	Nevada County	501,028	70,143 92	14		Official Rec., 1878	
Blue Lead	Nevada County	235,703	16,599 21	07	160–220	Official Rec., 1878	
Enterprise	Nevada County	1,308,963	28,818 63	02.0	150–300	Offici'l Rec.,1876-1878	Chiefly top gravel; fine, loose, and sandy material. W. H. Pettee. See "Contributions to Am. Geology," Whitney, p. 414.
Polar Star	Placer County	888,889	100,000 00	11		Colgrove Pettee	W. H. Pettee. See "Contributions to Am. Geology," Whitney, p.425.
Quaker Hill	Placer County	338,888		0⅗			
Kansas	French Corral, Nevada Co.	67,500	223,000 00	3 30	27	H. Smith, Jr.	Cement claim. The richest gravel selected and milled.
Empire	French Corral, Nevada Co.	29,166	200,000 00	6 85	18	H. Smith, Jr.	
Nebraska	French Corral, Nevada Co.	(Tons) 600	9,000 00	*		H. Smith, Jr.	See "Contributions to Am. Geology," Vol. I, Whitney, p. 425; also, Raymond's Rep, 1875, p. 100.
Indiana Hill	Placer County	14,449	75,422 47	5 22		W. H. Pettee	
Pond	Placer County	255,933	91,828 30	35.8	50	W. A. Goodyear	"Contributions to Am. Geology,"Vol.I,Whitney, p.118.
Sailor's Union	Placer County	404,615	42,800 00	10.5	116	W. A. Goodyear	The bedrock yielded $124,598. "Contributions to Am. Geology," Vol.I,Whitney, p. 116.
Paragon	Placer County	124,000	92,090 00	74.2	70	Jos. McGillivray	
Paragon	Placer County	22,275	17,387 78	78	71	Jos. McGillivray	
Dard'nell's and Oro	Forest Hill, Placer County	3,630,000	476,000 00	13.1	150	Jos. McGillivray	This deposit contained many large bowlders.
La Porte	La Porte, Plumas County		87,000 00	3 13	30	Chas. Hendel	
La Porte	La Porte, Plumas County		57,500 00	20 87	6	Chas. Hendel	Drifted.
Secret Diggings	Plumas County		300,000 00	2 00	140	W. H. Pettee	

Claim	County	Cubic yards	Value			Authority	Remarks
Gardner's Point	Plumas County	148,148	118,000 00	79	80	A. J. Bowie, Jr.	
Bean's Hill	Plumas County	314	220 00	70	5	A. J. Bowie, Jr.	} Shallow spots.
Jack's Hill	Plumas County	740	37 37	05.4	8	A. J. Bowie, Jr.	
McDoran's	Plumas County	5,555	300 00	61	20	A. J. Bowie, Jr.	
French Hill	La Grange, Stanislaus Co.	16,368	9,782 98	11.4	18	R. Abbey	
Light Claim	La Grange, Stanislaus Co.	73,566	8,468 35	2 83	57	R. Abbey	
Bald Mountain Co.	Sierra County	115,950	328,352 38			Raymond's Rep, 75.	Deep placer-mining. Cubic yards estimated from coarse dirt washed.
Bald Mountain Co.	Sierra County	50,040	286,737 28	5 93		H. W. Wallace	From June 24, 1875, to June 24, 1876, 100,080 carloads of gravel mined, stripping 300,240 superficial feet of bedrock. Each car contained one cubic yard of loose gravel, equal to one half cubic yard of gravel in place. Year ending June 24, 1877, 98,044 carloads extracted, stripping 343,768 superficial feet of bedrock. 1877-1878, 106,160 carloads extracted. 1878-1879, 90,274 carloads extracted. One thousand one hundred and eighty-six feet of tunnel in gravel, ten to twenty feet above bedrock.
Bald Mountain Co.	Sierra County	49,022	235,797 87	4 79		H. W. Wallace	
Bald Mountain Co.	Sierra County		289,755 00			H. C. Perkins	
Bald Mountain Co.	Sierra County		164,904 00			H. C. Perkins	
Pioneer Tunnel	Sierra County	883	1,400 53	1 59		Chas. Hendel	
E. Gr'nd Dry Creek	Shasta County	50,000	9,000 00	18		Raymond's Rep, '74	
W. Side Dry Creek	Shasta County	2,000	741 26	37		Raymond's Rep, '74	Raymond's Report, 1874.
Piety Hill	Shasta County	1,333	22,000 00	16 50	20	W. K. Conger	
Dry Creek	Shasta County	200 lin. ft.	170,000 00			Cooper	
Smartsville	Sucker Flat, Yuba County	2,042,880	400,000 00	19.5	112	Amos Bowman	
Union Gravel	Empire Hill	792,000	120,000 00	15	90	Amos Bowman	
Pactolus	Yuba County	1,468,800	295,000 00	20.8	85	Amos Bowman	See Report on the Smartsville Blue Gravel and Excelsior Canal Co., pp. 32-35.
Blue Gravel	Temperance Hill, Yuba Co.	2,419,120	1,560,000 00	63	83	Amos Bowman	
Pittsburg	Temperance Hill, Yuba Co.	565,700	237,000 00	41	59	Amos Bowman	
Pactolus	Yuba County	60,000	29,600 00	44	50	W. Ashburner and J. D. Hague	

NOTE.—These compilations were made by A. J. Bowie, Jr., E.M.

* $15 per ton.

Details of Work at No. 8 Claim, North Bloomfield Company.

	1874—1875.			1875—1876.			1876—1877.		
	Total.	Per Cubic Yard.	Per Inch of Water.	Total.	Per Cubic Yard.	Per Inch of Water.	Total.	Per Cubic Yard.	Per Inch of Water.
Cubic yards of gravel moved	1,858,000		4.8	2,919,700		4.17	2,293,900		3.86
Yield	$74,271 77	3.99 cents	19.19 cents	$192,735 73	6.60 cents	27.53 cents	$290,775 42	12.68 cents	48.87 cents
Expenses—Labor	$22,790 39	1.23 cents	5.89 cents	$40,975 85	1.40 cents	5.85 cents	$53,742 78	2.34 cents	9.03 cents
Explosives	2,944 94	0.16 cents	0.76 cents	10,279 73	0.35 cents	1.47 cents	25,376 16	1.11 cents	4.26 cents
Blocks	3,007 26	0.16 cents	0.78 cents	5,212 62	0.18 cents	0.75 cents	5,750 43	0.25 cents	0.97 cents
Material	5,643 89	0.30 cents	1.46 cents	9,250 46	0.32 cents	1.32 cents	10,158 72	0.44 cents	1.71 cents
Water	14,480 40	0.78 cents	3.74 cents	21,740 97	0.75 cents	3.11 cents	21,745 88	0.95 cents	3.66 cents
General	4,201 95	0.23 cents	1.09 cents	7,364 12	0.25 cents	1.05 cents	25,266 11	1.10 cents	4.25 cents
Totals	$53,088 83	2.86 cents	13.72 cents	$94,823 75	3.25 cents	13.55 cents	$112,060 08	6.19 cents	23.88 cents
Days run	295; com. Jan. 1, end. Oct. 14.			342; com. Nov. 13, end. Oct.18.			318; com. Nov. 2¾, end. Oct. 13.		
Grade of sluices	6¾ inches to 12 feet.			6¾ inches to 12 feet.			6¾ inches to 12 feet.		
Height of banks	180 feet.			200 feet.			318 feet.		
Inches of water	386,972.			700,000.			595,000.		

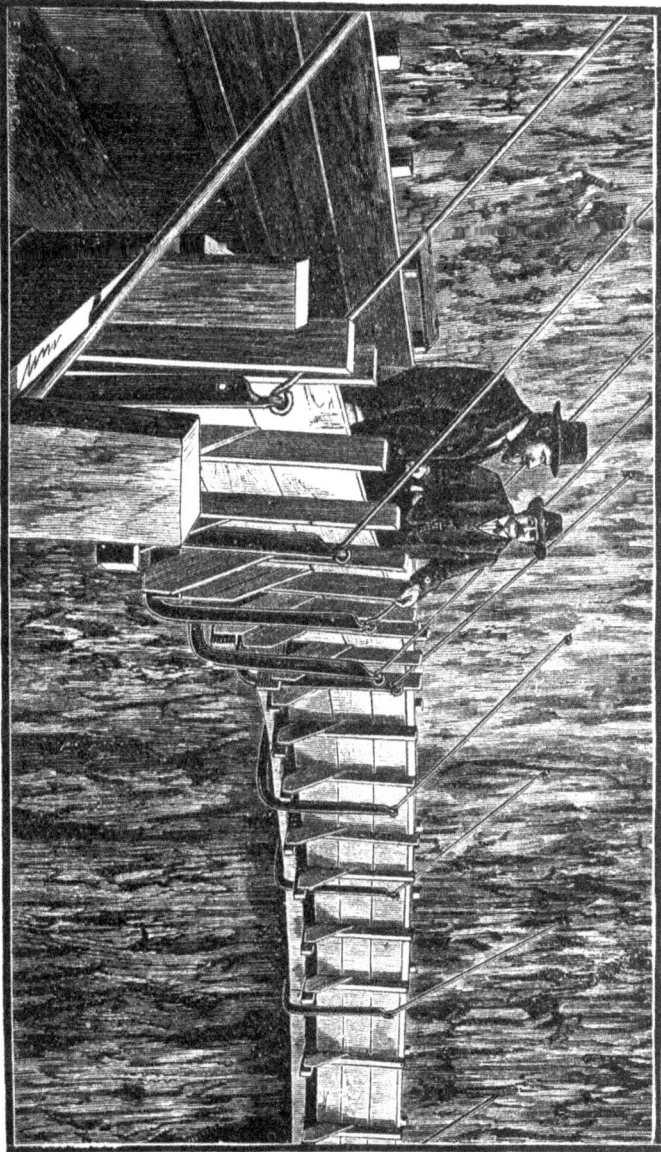

MIOCENE COMPANY'S BRACKET FLUME

www.ingramcontent.com/pod-product-compliance
Lightning Source LLC
Chambersburg PA
CBHW071236200326
41521CB00009B/1493